# INTERNATIONAL RESEARCH FORUM 2009

# INTERNATIONAL RESEARCH FORUM 2009

## LUTZ HEUSER, HENRIKE PAETZ, AND DAN WOODS

Evolved
Technologist
Press
New York, NY

**International Research Forum 2009**
Lutz Heuser, Henrike Paetz, and Dan Woods

Published by Evolved Technologist Press, an imprint of Evolved Media, 242 West 30th Street, Suite 801, New York, NY 10001

This book may be purchased for educational, business, or sales promotional use. For more information contact:

Evolved Technologist Press
(646) 827-2196
*info@EvolvedTechnologist.com*
*www.EvolvedTechnologist.com*

Editors: Dan Woods and Deb Cameron
Writers: Dan Woods and Kermit Pattison
Copyeditor: Deb Cameron
Production Editor: Deb Gabriel
Cover and Interior Design: 1106 Design
Illustrator: Deb Gabriel
Photographs of Lutz Heuser and Henrike Paetz: Stephan Daub
First Edition: April 2010

ISBN: 9780982550601; 098255060X

# Contents

n April 2009, leaders from business and academia gathered for the fourth **SAP International Research Forum** in Dresden. The participants represented some of the leading lights in the world of business and technology and, over two days, they discussed, debated, and questioned how the Internet of Things (IOT) would play out in the years ahead. Four sessions explored key topics facing the Internet of Things: end-to-end real-world awareness, infrastructure, potential killer applications, and future manufacturing. Participants listened, added trenchant observations, parried in lively intellectual debates, and posed provocative questions.

Given the time constraints, many questions were left unanswered. The task then fell to the authors of this book and their research staff to follow up on these questions. They conducted months of additional research and interviewed more world-leading experts. The result is the book you are reading right now.

The first half of this book summarizes the conference proceedings. The second half examines how the IOT will be built, how it will be used, and how it will grow. This quest takes us to the frontiers of the digital and physical worlds.

Please join us on this journey.

— Lutz Heuser

## Acknowledgments

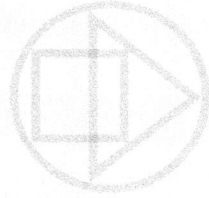

No project of this scope could be accomplished without the generous assistance of many friends and colleagues.

First, we must thank the participants of the **International Research Forum 2009.** They served as the brain trust that generated many of the observations and questions discussed throughout this book. During the months of research that followed the forum, the authors of this book were amazed at how many times our investigation echoed or confirmed comments made by participants at the event. There are too many people to list here; please see the appendix for a full listing of the participants. We are fortunate and grateful for their participation.

Too often, the insights produced at conferences dissipate at the end of the event. In contrast, the discussions at the IRF 2009 were captured thanks to the team at SAP Research who recorded it in video, audio, transcripts, and mind maps—all of which provided a rich record and excellent start for the writing team. Many thanks to Christiane Beck, Sabine Patsch, and to the rest of the SAP Research Dissemination team for their help organizing the event.

Moderators Dan Woods and Mark Jeffries kept the discussion moving along. They asked provocative questions, pressed for clarification, and helped draw out key issues that enriched the discussion.

We also are grateful for the many experts who shared their wisdom in interviews after the forum. They included many colleagues within SAP Research and other parts of SAP. We are fortunate to be surrounded by so many wise colleagues, but we also recognize that we have much to learn from those outside our company. We are very grateful to all the experts who joined us in interviews and shared their insights. Conducting these interviews was a delightful task. The interviewees who were not at the Forum itself are listed in the Appendix as virtual participants, and we thank them one and all.

The writing team of Dan Woods and Kermit Pattison turned the raw transcripts into an engaging narrative. Transcriptionist Richard Fisher produced fast and accurate records of our conversations. Project manager Deb Gabriel wrangled interviews, edited countless revisions, and ably performed the monumental task of keeping all the parts of this project on track and on schedule. Managing editor Deb Cameron improved the manuscript in countless ways with her keen eye. The text was greatly enhanced by the visually-pleasing illustrations by Deb Gabriel and the cover and interior design by Michele DeFilippo and her team at 1106 Design.

As always, we are also deeply indebted to the SAP Executive Board for their vision and support of this project.

IRF 2009

Part One

Part One describes the proceedings at the 2009 International Research Forum. This part represents reporting of the conference as it occurred. It captures the presentations by the main speakers and discussions among the participants. These proceedings laid the groundwork for and framed the questions about the research that shaped the second half of this book.

# 1

The Internet of Things (IOT) rests on a simple idea with profound implications: that virtually every physical thing in the world can be connected to the Internet. These items may be tagged with small computers, miniature devices like Radio Frequency Identification (RFID), or simply a digital address. In this vision, almost every physical object can be described digitally and read by machines. The IOT merges the digital world with the physical world.

The idea of the IOT has been around for many years. But this vision has moved closer to reality as technological advances reduced the size, cost, and energy consumption of small devices. It is becoming feasible to manufacture and deploy great numbers of small, low-power, and inexpensive devices like RFID tags, sensors, or tiny computers.

## Setting the Stage: A Brief Primer on the Internet of Things

The IOT goes by many names, such as pervasive computing or ubiquitous computing. All capture the same basic idea—a globally interconnected network with millions or billions of small nodes permeating every aspect of the physical world. The IOT goes far beyond the embedded systems we see proliferating in our world today. Embedded systems are becoming more commonplace in factories and devices; 90% of processors now go into devices other than traditional laptop or desktop computers. But IOT should not be confused with these embedded systems, which are often isolated with their own protocols and buses. The vision for the IOT represents a network of interconnected devices that communicate via the Internet.

These devices act as the nerve endings of this global nervous system. They may be passive devices like RFID that store and disclose data when queried. They may be more advanced technologies like active RFID that broadcast this information and communicate with other devices. They may be cameras or sensors that record data about their surroundings, such as temperature, location, speed, or humidity. At the higher end, they may be miniature computers with local processing power that sense the world around them and decide on a course of action without being prompted by humans. One example is Mr. SemProM, a robot warehouse worker who we will meet later in this book.

This vision for the IOT carries far-reaching implications. As we will see throughout this book, it will bring fundamental changes to business processes, business models, and business relationships. In the not too distant future, food packages on the grocery shelf will carry tiny RFID tags with the life history of every ingredient and a diary of its entire life from farm to freezer. Smart meters may transform how we heat, cool, and light our homes. Machines will automate factories and turn even wrench-wielding workers on the assembly line into information workers. Indeed, a personal device may become as *de rigueur* to the laborer as a union membership. These technologies will transform entire sectors of the economy, such as manufacturing, supply chains, energy production, distribution, and many more. It will alter how businesses provision services and how people consume them. It will transform the nature of management itself: this explosion of data will allow managers

## Key Differences

In many ways, the IOT may be seen as an extension of the Internet. After all, it rests on Internet building blocks, such as the Domain Name Systems (DNS) and Transmission Control Protocol Internet Protocol (TCP/IP). At the same time, however, it is a distinctly new phenomenon with unique requirements and characteristics. Before plunging into the subject, it is worthwhile to briefly sketch out some key features. Prof. Dr. Elgar Fleisch, Professor of Information and Technology Management, ETH Zurich and University of St. Gallen, has identified six key differences between the IOT and the standard Internet:

- **Unflashy hardware:** In the conventional Internet, the nerve ends are full-blown computers, such as workstations, laptops, or mobile phones. In the IOT, they are small, even invisible, with very limited functionality. They are limited by size, low power, and connectivity. In most cases they cannot communicate directly with people

- **Trillions of nodes:** In the IOT, the numbers of devices will increase by orders of magnitude. In the Internet, the number of network nodes is a matter of billions; in the IOT, it will be one of trillions. There will be so many computer-enabled objects that people will not be able or willing to directly communicate with them all. As a result, some kind of new network infrastructure might be required

- **The last mile:** In the Internet, people talk about turning the proverbial "last mile" into an information superhighway. In the IOT, the last mile can be the eye of a needle. The IOT requires considerably less bandwidth per device. The speed of the last mile of an average RFID tag is only about 100 kBit/s, far below the broadband connection of the average household. Yet the last mile presents another challenge: bridging the air gap between the sensor and Internet. It must be wireless, durable, energy efficient, and secure

- **Global address routing:** The Internet-based addressing via MAC and IP schemes today requires too much capacity to be viable in the IOT. We need global standards for bridging the last mile from the Internet to the smart devices and better technologies and standards for addressing. So far, most identifiers use closed-loop schemes from specific vendors or industries

- **Machine-centric:** In the Internet, most services are designed for human users. In the IOT, machines talk to other machines and often exclude human involvement

- **Sensing more than communication:** The Internet was about communication with people. The IOT is mostly about sensing the physical world

to engage in the new art and science of "high-resolution management," where information and action are compressed into a near-instant loop.

The IOT represents a significant force in the business world. Forrester Research predicts that the total market for IOT technologies will sustain a growth rate of 48% and exceed US $11.5 billion by 2012. ABI Research is even more bullish: it predicts a market volume of more than US $27 billion by 2011.

Yet, as we shall see later in this book, much work remains to realize this vision. This is not simply a matter of adapting existing technology. Rather, the IOT is likely to require new standards, new infrastructures, new architectures, new software, new methods of data management, new methods to preserve security and privacy, new business models, and new methods for sharing costs and benefits.

## The International Research Forum 2009

The **International Research Forum 2009** got underway on a sunny spring day in Dresden. In his welcoming remarks, Prof. Dr. Lutz Heuser, Executive Vice President and Head of SAP Research, explained that the location in itself was significant for more than the gorgeous view outside the windows of the conference room. Dresden is the center of many SAP research activities. Later, the group would take a field trip to see one of SAP's Living Labs—and thus bring some of the conceptual ideas from the conference down to earth with a real-world example of practical applications.

Prof. Heuser noted that the fourth IRF built upon previous conferences and once again would become a forum for a lively exchange of ideas. With that, Prof. Heuser launched into the first presentation of the day, which he called the warmup session for the rest of the conference. He summarized the previous year's IRF. In brief, the infrastructure barriers to doing business in the Internet of Services are dropping rapidly. New technological trends, such as software as a service and cloud computing, are democratizing innovation. As a result, there is massive opportunity for new innovations and killer apps, especially in the burgeoning field of mobile technology. All these topics led straight to this year's destination—the IOT.

But the IRF always has stimulated much more than talk. Prof. Heuser summarized highlights of the initiatives born out of the conferences. Some of these findings were incorporated into the THESEUS-TEXO project, a collaborative effort between SAP and other partners, such as the German government, Siemens, and the DKFI, the German Research Center for Artificial Intelligence. "And last year, stimulated by the activities here at IRF and other places, we could show that there is a web-based service industry to come," he said.

The THESEUS-TEXO project explores potential killer applications. It has developed models for a service marketplace for enterprise services and even mobile access to services. It was featured in a demonstration at the CeBIT 2009 conference. TEXO researchers also have developed an advanced semantic service discovery model, a combination of structured and unstructured search for standardized services. They also have introduced USDL (Unified Service Description Language), a standardization language for description of Internet services.

The IRF community has launched several new initiatives. In Australia, SAP has teamed up with Australia's Information and Communications Technology ICT Centre of Excellence (NICTA) and Fraunhofer IESE to set up a Living Lab. The research lab will focus on visualizing the future of Transport & Logistics networks and will involve further local and global players. Researchers and industry will work and disseminate results from the Internet of Services and the Internet of Things (IOT) to create an Internet of Transport & Logistics, immersing visitors in a holistic, sustainable, and dynamic end-to-end scenario ranging from manufacturer and different modes of transportation to consumer. Goods, processes and information flow will be made transparent, enabling optimized decision-making while minimizing cost.

In Germany, SAP and its partners including IDS Scheer, Software AG, and the DKFI have been busy on a project that also grew out of the last IRF. They have formed an alliance to improve the flow of digital information that accompanies the flow of goods and explores how to build better business processes that leverage the IOT and Internet of Services (project ADiWa—a German abbreviation for "Alliance for digital product flow"). The group is examining how to combine the Internet of Services with the IOT to create more agile end-to-end business processes. Even as the IRF 2009 got underway, a group of 60 other researchers gathered in Walldorf, Germany, for the ADiWa project. This project is not only an outgrowth of the IRF, noted Prof. Heuser, but also part of a bigger picture of related projects. In particular, it leverages two other important projects: Aletheia and SemProM. (We will discuss these projects in greater detail later in this book.)

The Aletheia project employs semantic technologies to enrich every phase of the product lifecycle with more information. It uses semantic technologies to integrate heterogeneous sources of information, ranging from the IOT, to business content, to Web 2.0 user content. It already has made significant advances in product memory. "All of that now has formed probably one of the largest clusters in the world dealing with the integration of the Internet of Services and the Internet of Things," said Prof. Heuser.

SemProM explores semantic product memory (also known as digital product memory). It researches how technologies like RFID and embedded systems can be used to store information directly on products and accompany the items through their entire lifecycle. The project has identified 52 use cases in areas such as manufacturing, logistics, and retail. These include carbon footprint tracking, decentralized manufacturing controls, and auto parts tagged with their own maintenance records. This project is led by Prof. Dr. Wolfgang Wahlster, Director and CEO of DFKI, the German Research Center for Artificial Intelligence, and Professor of Computer Science at Saarland University.

In short, the various initiatives span the gamut from simple barcodes to embedded systems with built-in intelligence. These kinds of technologies, said Prof. Heuser, will fundamentally change the architecture of Enterprise Resource Planning (ERP), especially as organizations increasingly empower the edges of their organizations and move away from centralized systems.

Last year, one of the sessions of the IRF 2008 explored killer apps. Later that year, one such killer app appeared on the market for emergency response and disaster management. The SoKNOS project application shortens the inevitable chaos phase in the early stages of disaster response. It ensures a continuous flow of information from different data sources and enhances collaboration among different players, such as police, fire, and other emergency services, as they operate under time pressure. The

Future Public Security Center is one of four current Living Labs at SAP Research and SoKNOS is one of its main projects.

Imagine a flood in the city of Cologne. This program would provide a dashboard with webcam images, satellite photos with danger spots marked in red, forecasts about water levels, and display messages and information about the whereabouts and status of emergency crews. It would provide a real-time collaboration platform for the organizations involved in the disaster management.

But luckily on this day, the weather was pleasant and the only floods were of information.

Turning to this year's conference, Prof. Heuser pointed out that the "Future Internet" consists of two equally important components, the IOT and the Internet of Services. Key elements and application areas include high-resolution management, digital product memory, future transportation and logistics, future manufacturing, e-energy, public security, and personalized financial services.

"The Internet of Things is our main topic today, but it relates to many other topics," said Prof. Heuser. "We should not be religious and single out the IOT, but take a broader view considering other elements as well."

Prof. Heuser outlined one such broad view of the Future Internet. In this vision, the IOT will form part of the basic infrastructure and foundation of the Future Internet. On top of this infrastructure will be layers of Internet services and a future web-based service industry. This vision is shared by the European Commission's Information Society Technology department (IST) and reflects thinking among leaders across the continent.

This notion raises several additional questions that need to be addressed: semantic technology, embedded systems, integrated solutions, sensor network technology, and service-oriented enterprise applications. All of these elements are intertwined with the IOT. Prof. Heuser noted that the topic should emphasize the first part of the name—Internet. These devices realize their potential as part of networks. "It's not the smart things alone," he said. "It's about how are they going to communicate among themselves as well as with the other systems on the Internet."

He offered a sneak preview of key topics that would resonate throughout the two-day forum, and in the months of subsequent research by the authors of this book. How will the business world make the transition from centralized processing to more distributed and localized decision making? What does it mean to be in charge of a business process that you can no longer control? How will high-resolution management change business processes?

The IOT will unleash far-reaching changes—ones that go far beyond the actual devices. It will lead to not only product innovation, said Prof. Heuser, but also to service innovation and business model innovation.

"We should discuss all three," said Prof. Heuser. "If we leave here and only talk about product innovation, we will definitely fall short of what needs to be done."

Indeed, there was much to be done. Over the next two days, IRF participants explored four cornerstone topics of the IOT: end-to-end real-world awareness, infrastructure, killer applications, and future manufacturing. They engaged in lively debates and spirited discussions.

It was fully appropriate that Dr. Uwe Kubach, Director of the SAP Research Center Dresden, was entrusted with the task of giving the first vision speech of the Forum. Dr. Kubach is spearheading SAP's efforts in Future Manufacturing. Later that evening, he would invite his fellow participants on a tour of the Future Factory, described in Chapter 5.

## The Vision

Dr. Kubach began by showing how last year's topic, the Internet of Services, led naturally to this year's topic of the Internet of Things. Together, these two trends create new systems of intelligent businesses processes.

"End-to-end real-world awareness is basically about merging the Internet of Things and the Internet of Services for more customer relevance," he said.

**Dr. Uwe Kubach** is the Director of the SAP Research Center Dresden. Among other activities the Center is driving SAP's research in the fields of future manufacturing, smart items, and data management and analytics. Dr. Kubach initiated a number of research projects in these domains and regularly acts as an industrial consultant to organizations such as the European Commission.

Dr. Kubach has a background in compiler engineering and distributed systems and earned a Ph.D. from the University of Stuttgart and an Executive M.B.A. from the University of Mannheim.

Two major technological trends are at work. First, we have improving abilities for sensing and computation on smart devices. Second, we have ever-expanding communication and connectivity.

RFID represents an early phase of those trends, but is only the beginning. Dr. Kubach predicted that we will see ever-greater levels of real-world awareness, such as wireless networks, embedded systems, and other devices that talk to each other and make decisions. These have given rise to collaborative systems with peer-to-peer communication and built-in business logic.

One example is a recently completed SAP project for self-monitoring chemical drums. These drums are equipped with sensor nodes and are able to independently monitor themselves, talk to each other, and keep track of storage regulations. The system can detect when they exceed volume limits, confirm safe storage environments, and trigger alerts to prevent unsafe storage combinations.

But these advances have yet to capitalize on many other business opportunities. "We only used IT and smart items for process optimization and cost reductions within a company," said Dr. Kubach. "What we missed was collaboration across companies to bring this value to end consumers."

Happily, we now have tools to close this gap.

The missing piece is the Internet of Services. These smart devices, combined with web services and service architectures, allow us to create

Applications make use of the Internet of Things and the Internet of Services to provide access to information and automate processes

**Applications**

*Makes Use of IOT*

*Makes Use of IOS*

**Internet of Things**

Make Use of Each Other

**Internet of Services**

The Internet of Things reaches into the physical world and collects information and events based on a distributed network of sensors, processors, and the ability to identify objects. The Internet of Things provides this information to applications and to specific services in the Internet of Services

The Internet of Services provides access to a huge range of functions that can be used by applications or by the Internet of Things. The Internet of Services can benefit from the information provided by the Internet of Things

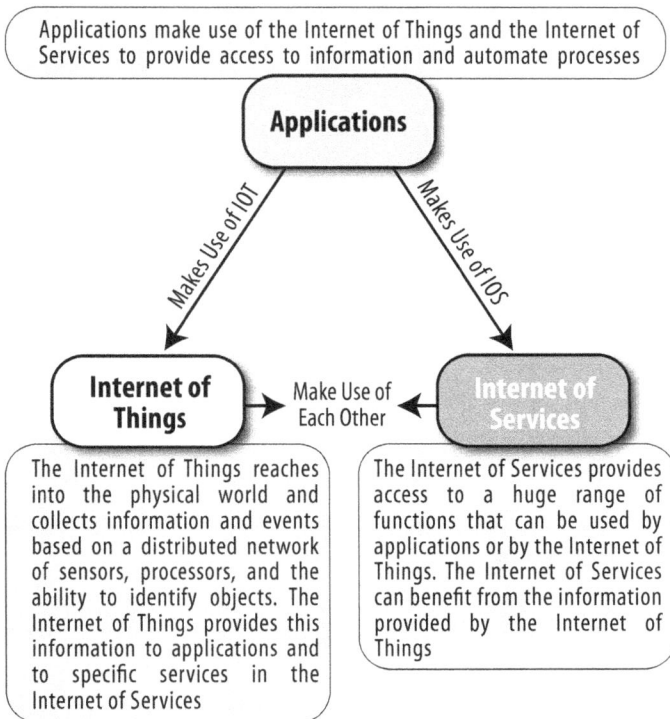

*Figure 2-1. The Internet of Things and the Internet of Services*

flexible, lightweight processes that can be quickly reconfigured in a completely new context. "With that," Dr. Kubach said, "we can go a step further and really bring value to the end users."

Some of these systems already are well underway in business-to-business applications and collaborative networks. The next step, said Dr. Kubach, will be mashups that allow users to combine and consume services from these intelligent objects on the IOT.

The Internet of Services gives users flexibility, a service platform, and the ability to easily assemble services and put them to work right away. The IOT provides tools for efficient data collection, real-time awareness, local intelligence, and self-executing business processes.

In the future, businesses will increasingly expose information from the IOT through software or software as a service. Until now, these combined capabilities have been leveraged mostly in internal logistics processes within certain companies. As they mature, they will increasingly be shared with business partners, customers, and others. Dr. Kubach said, "Then we'll really have a new value to offer end consumers."

But there is no substitute for seeing these processes in action. That night, Dr. Kubach invited his colleagues to his research center, where they were guided through the Future Factory. The Future Factory Initiative is pushing forward the frontiers of end-to-end real-world awareness in manufacturing. It conducts research and development, incubates B2B collaborations between partners, and acts as a showroom for innovations.

"The Internet of Services and the Internet of Things together really provide means for an end-to-end real-world awareness," Dr. Kubach concluded. "This offers value for business partners and the end consumer."

## Another Perspective

Another perspective of the IOT came from Prof. Dr. Ryo Imura, founder, President, and CEO of Mu-Solutions Company in Hitachi Ltd. He described a vision for a "ubiquitous world." In such a world, people expect secure and easy access to network systems from anywhere, at anytime. They also expect to get information about things, such as data and location. This requires a system of traceability and a network where objects are able to talk to each other.

In fact, one of the killer apps of the IOT is the network itself. RFID is the preeminent technology currently in wide use. It was the promise of this technology that inspired Prof. Imura to establish his company.

With RFID, smart systems can track products, confirm safety, quality, and reliability, protect the brand image, and allow tracing forward and backward. There are many successful applications of this technology, including food traceability, smart libraries, e-tickets, keeping tabs on equipment for maintenance, tracking shipments, restricting access to apartment buildings, and preventing counterfeiting.

Yet this promising technology needs to surmount some major barriers. According to Prof. Imura, RFID needs better public understanding and acceptance. It needs more innovative value chains involving multiple partners. It must show practical applications with clearer benefits and returns on investments. Solutions must address concerns such as privacy. With this brief list, Prof. Imura foreshadowed challenges that will surface again and again in the IOT.

He pointed out one problem with what is perhaps the most famous example of an RFID system: Wal-Mart. The system has worked splendidly for Wal-Mart—but not necessarily for its partners. Wal-Mart mandated RFID tags to help the retail giant manage its supply chain and inventory. But the suppliers, who bear much of the cost, see fewer benefits.

"We need some mechanism for cost sharing and the balance of cost and benefit," said Prof. Imura. "That's why Wal-Mart faced tough difficulties that delayed the establishment of their business model."

Instead, we should create networks that benefit all partners in the value chain. Such a ubiquitous network must balance the needs of industry, society, and the individual.

To gain widespread public acceptance, these systems must demonstrate clear benefits to society through practical applications. They must show continuous innovation. They must establish a reliable network system and management system for the entire product lifecycle.

## The Challenge

There was much to admire in this vision. Yet this scenario also raised many questions that remain unanswered. The task of listing these challenges fell to Prof. Dr. Hubert Österle, Professor of Business Engineering and Director of the Institute of Information Management of the University of St. Gallen, Switzerland.

"We have seen very great visions and great projects," he began. "And so I'm going to disturb this picture a little bit."

Indeed he did. Prof. Österle has seen many of these shortcomings up close—and even has a burglarized house to show for it. He and his colleagues at the University of St. Gallen worked on a project called Independent Living, which tries to leverage the technologies of the IOT to create a more intelligent environment for elderly people. They studied homes with sensors and intelligent devices intended to make life easier for the elderly—an environment known as "ambient assisted living."

They soon learned a sobering lesson.

**Prof. Dr. Hubert Österle** is Professor of Business Engineering and Director of the Institute of Information Management of the University of St. Gallen, Switzerland, since 1980. He was founder and executive board member of the Information Management Group (IMG AG) from 1989 until 2007. Currently the Editor-in-Chief of *Electronic Markets – The International Journal on Networked Business,* Prof. Österle is also the author of numerous books and other scientific publications and a member of several scientific and industry boards.

Prof. Österle's research is focused on three areas within business engineering: corporate data quality, independent living, and sourcing in the financial industry.

The sensors that were supposed to detect movement or monitor heart function and sleeping didn't work. Neither did another "electronic pillbox" gadget that was supposed to dole out medications at prescribed intervals.

"The European Union spent millions and millions of euros in projects for ambient assisted living," said Prof. Österle, "but we don't see real working environments where elderly people can really be assisted and not hindered."

Simply put, the systems weren't up to the task.

Neither are many other so-called systems of the future. Prof. Österle illustrated with another personal example. As a self-described gadget addict, when he built his own house, he wired it with a security system to protect his family. Soon, he awoke at 3 A.M. with the alarm blaring and lights flashing. He looked around the house, found no burglar, and fired up his computer to diagnose the problem. A sensor indicated an open door—except that the door really was closed. The sensor was broken.

The alarm went off another time. Prof. Österle got an urgent call from a neighbor, cut short a work meeting, rushed home, and found another false alarm—this time due to a faulty motion detector.

Eventually he stopped using the alarm system. Then one night he came home late and found his doors open. This time there really had been burglars—and they had gotten inside undetected because the alarm was unarmed.

"You will ask, why didn't you use your alarm system?" he said. "Because it's too complicated. We had too many problems with this alarm system so we didn't put it on that time."

Moral of the story: these systems aren't yet reliable.

"I know of almost no installations in the consumer environment that are really working," he said. "Parts of it do, but if you want an integrated system, as I do, it doesn't work."

The problem isn't technology *per se*. Prof. Österle recalled a recent conversation with the CEO of a tech company who received 200 calls per week from providers who wanted to sell him new technology. They already had plenty of technology. Instead, they needed systems and processes. Additionally, they needed new business models making the use of these systems attractive to vendors and consumers.

Thus, the IOT faces two basic challenges. First, in many cases, the technology isn't ready for prime time. The vision is ahead of the implementation. It faces all the typical shortcomings of early-stage technologies: it's expensive to install, it's complicated to handle, and it lacks sufficient standardization, security, and scalability.

"The second and even bigger problem is the business model," Prof. Österle said. "We don't just need the products, we need the services around the products that make the products usable."

A technical model and a business model are two different animals. A technical model, said Prof. Österle, includes the four basic elements of usability, standardization, security, and scalability. A business model has more requirements: services and products, added value for customers, sales channels, specialized skills, business processes, ecosystems, and revenue and cost models.

When it comes to fulfilling these requirements for the business model, the IOT still falls short of the vision. We don't know the added value for the consumer, and we need new sales channels. We need new skill sets. We have to learn the processes of the consumers and tailor the solutions to their needs.

"We have to build the ecosystems, companies working together so they are able to provide all the service and technology that is necessary to install a solution for the consumer," Prof. Österle said. "And finally, we need revenue and cost models."

Business models will be the driver for the IOT, yet these models need to mature before the full potential can be realized. As we will see later in this book, we need to resolve many related issues: usability, technological integration, standards, new architectures and techniques for data management, fair mechanisms to share costs and benefits, and useful applications.

Prof. Österle concluded with one final bit of advice: "If you are going to plan an electronic house, give me a call first."

## The Discussion

The participants began their first breakout sessions and panel discussions. Participants divided into four groups to further explore some of the issues raised in the vision and challenge speeches. What prospects for the IOT appeared promising? What seemed to be just a chimera? After the participants reemerged from their breakout sessions, a spokesperson summarized the findings of each group. These experts quickly surfaced many of the key questions facing the IOT and their observations offered a snapshot preview of the topics that will resonate throughout this book.

Dr. Matthias Kaiserswerth, Director of the IBM Zurich Research Laboratory, questioned the scalability of the IOT. Thus far, successful applications have been small experiments in local settings. What happens when we try to expand this technology to the real world and create more complex linkages with many players?

"We know about the feebleness of electronics; we still need to reboot our PCs every once in a while," Dr. Kaiserswerth said. "Multiply this by hundreds of thousands of sensors and actuators and suddenly the reliability of the system goes down to zero."

IOT systems need a layer of intelligence. Dr. Kaiserswerth suggested that we need an "information management layer" on top of the machines that makes intelligent decisions. Is there really a burglar or just a faulty sensor?

Prof. Dr. Otthein Herzog of the University of Bremen voiced a similar concern. "We have to make sense of sensor data," he said. "We have sensor technologies and we can get huge amounts of data, but we don't know exactly whether the data is right or not."

What role should humans play in these mechanical systems?

"We came to the conclusion that the more the human being can participate, the better the solution is likely to be," said Dr. Nelson Mattos of Google.

Dr. Mattos said the IOT must benefit all participants. If not, they will be unlikely to participate and the vision will be stymied. Yet the ideal of collaboration raises a host of thorny problems.

"How do you guarantee access to all the participants?" asked Dr. Mattos. "What level of the data are they allowed to access? How can we unify semantics coming from siloed applications, industry standards, and so forth?"

Concerns about privacy, security, and trust were echoed by many speakers. Mary Murphy-Hoye, Senior Principal Engineer in Intel's Embedded & Communications Group, took a pragmatic approach: we should assume we don't have any privacy and security and go from there. We need to move from defining security with a fortress mentality, since the Internet has made it impossible to build walls around an enterprise and its data. Instead privacy becomes personal and context-dependent and information and things must carry self-defining security attributes.

Prof. Dr. Oliver Günther, Dean of the School of Business and Economics at Humboldt University, Berlin, Germany, took another tack: we should look at privacy as an asset that we can trade.

"With cross-company cooperation, you see exactly the same things," he said. "Companies are sharing information willingly if they get something in return, so there's a little ROI calculator in our brains and privacy is just a currency that we're using."

This brought the discussion back to the widely expressed disappointment with Wal-Mart's RFID model. According to Prof. Günther, one reason for its failure was that there was an unequal exchange of currency; the participants didn't feel good about it.

"It's about creating value for most, if not all, of the participants," he said. "And the currency is not really money—it's long-term potential. It can be strategic advantages, it can be data, it can be privacy. All these are currencies."

What about people? Prof. Herzog asked, "Does end-to-end real-world awareness also mean awareness of the social environment?"

More questions followed. How can we begin to define use cases for this end-to-end real-world awareness? What has been the impact of end-to-end real-world awareness in business processes? And how do we incorporate technology questions, business questions, and human factor questions?

Prof. Dr. Pradeep Khosla, Dean of the College of Engineering at Carnegie Mellon, questioned the very definition of the IOT.

"To me, it should be anything and everything," he said. "The Internet of Things should be about entities that have an IP address, period. Whether they are services, hardware devices, whatever, human beings, it doesn't matter."

Forum moderator Dan Woods, CTO and Editor of CITO Research, linked the IOT with a topic from a previous IRF known as real-world awareness and high-resolution management. This topic was explained by Prof. Fleisch, ETH Zurich and University of St. Gallen, in the first IRF. We will revisit high-resolution management and Prof. Fleisch extensively throughout this book. In a nutshell, high-resolution management is based on a simple insight: most business processes are constructed on the idea that information is expensive. But, asked Mr. Woods, what happens when the cost of this information drives toward zero? How would that change your business?

"If getting more information is very cheap—and that's what the Internet of Things provides in many ways—how would you change the way your business processes are executed?" asked Mr. Woods. "Right now, in most business processes, that rethinking hasn't been done. Most business processes are based on a world in which information is hard to come by."

Prof. Fleisch provided some insight into how we might use the IOT and high-resolution management to attack chaos and complexity. "But the resulting processes that you come up with are not going to look like the processes that we have today," said Mr. Woods. "They're going to be much more complicated, but we have the technology and the information to deal with them. In doing this, a bunch of issues are going to come up that are very testy—issues of privacy, issues of security, issues about who will be the designers of these systems."

The Forum generated so many ideas and questions that it was impossible to answer them all in only two days. The ideas were captured on boards where participants could jot down questions, ideas, web sites, experts to interview, and so on. This ensured that good ideas and provocative questions would not be lost as the Forum moved ahead at a fast pace. It also left the editorial team with a "to-do list" of items to follow up with additional research as they spent the next several months writing this book.

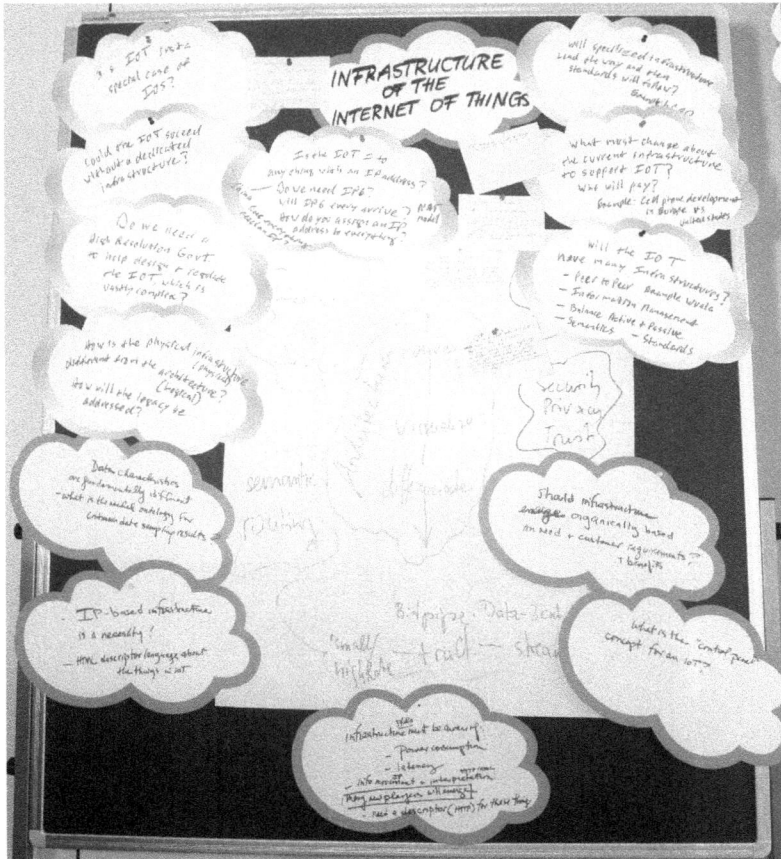

*Figure 2-2. The Idea Board*

By the end of the first day, many provocative questions had been thrown into the air. It was time to leave the conference room and see some of these ideas in action. With that, the group adjourned for a tour of the Future Factory Living Lab of SAP Research Center in Dresden.

What will be the infrastructure of information technology for the Internet of Things? The task of setting forth this vision fell to Prof. Dr. Max Mühlhäuser, Head of the Telecooperation Lab at Technische Universität Darmstadt, Informatics Department.

## The Vision

Prof. Mühlhäuser encouraged people not to get too discouraged by critics who point out shortcomings in the IOT. This, he said, is the "trough of disillusionment"—a typical part of the lifecycle of any much-hyped technological trend.

He envisioned the IOT as part of a larger ecosystem of the Future Internet. Other components include the Internet of Services, the Internet

**Prof. Dr. Max Mühlhäuser** is Head of the Telecooperation Lab at Technische Universität Darmstadt, Informatics Department The Lab works on smart ubiquitous computing environments for the 'pervasive Internet.' He also heads the 'RBG' division for e-Learning and computing services and is a Directorate member of CASED, a center for advanced security research. He is founder and speaker of a center of research excellence on e-Learning and of a corresponding graduate school funded by the National Funding Agency DFG. Prof. Mühlhäuser's academic and technology transfer appointments include chair of the academic steering committee at SAP Research, CEC Darmstadt lab.

Prof. Mühlhäuser has about 25 years of experience in research and teaching in areas related to ubiquitous computing (UC), networks and distributed systems, and e-Learning. He regularly publishes in *Ubiquitous and Distributed Computing*, HCI, Multimedia, and e-Learning conferences and journals. He is a member of editorial boards or guest editor in journals such as *Pervasive Computing, ACM Multimedia, Pervasive and Mobile Computing, Web Engineering*, and *Distance Learning Technology.*

of Crowds (social networks, crowdsourcing, Web 2.0, and virtual words), the Internet of Clouds (computing, storage, and services based on Cloud Computing), and the Internet of Humans (which provides secure and privacy-protecting global access for the individual).

"The Internet of Things is part of a family," he said, "so we should really look at the big picture."

How will this family mature? Prof. Mühlhäuser peered six to ten years into the future and made a few predictions. Armed with technologies like RFID, the enterprise will reach out in real time to the real world with all kinds of IOT technologies. This goes beyond the much-ballyhooed mobile technology. Rather it's the sort of technology that arms workers with intelligence—just as IRF participants saw the night before, when they visited the Future Factory.

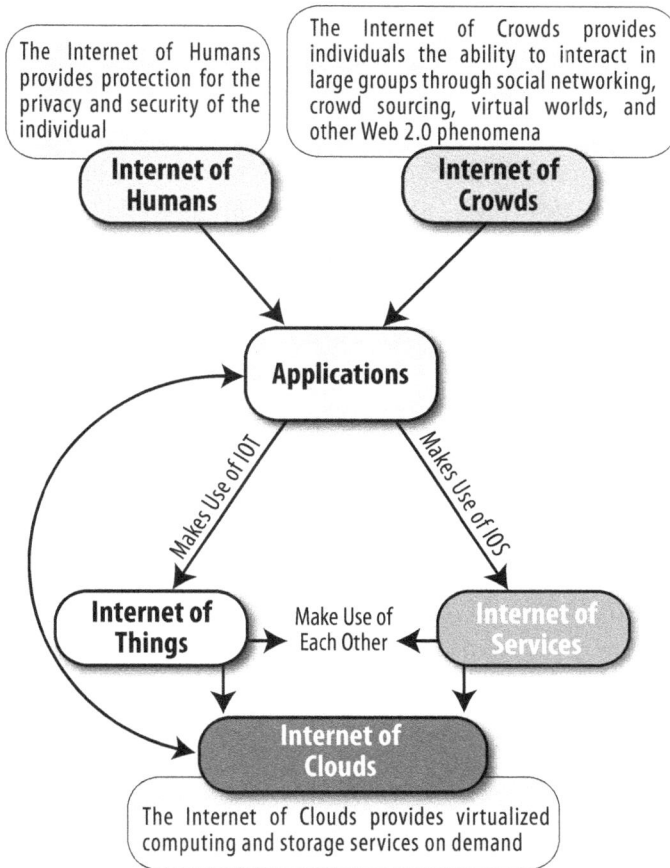

*Figure 3-1. The Future Internet*

"Ten years from now, we'll really have a closed loop between the humans out there in the processes, not at the desktop."

According to Prof. Mühlhäuser, the IOT faces three key challenges:

- It must bring humans into the loop

- It must solve issues of privacy, security, and trust—which are not only technical issues, but also emotional issues that will determine public acceptance of these technologies

- It must develop new, highly distributed and scalable architectures

## Humans in the Loop

At this point, we lack interaction support for the "wrench users"—the blue-collar workers on the front lines of manufacturing processes. These systems have thus far failed to incorporate the end consumers. Granted, there have been some efforts like WearIT@Work and other European Union projects. But these initiatives have failed to yield standards or engineering methodologies that allow widespread, low-cost development of processes that bring users into the loop.

"Both the blue-collar workers and the customers need to get involved and really integrated in the loop of what we are doing," said Prof. Mühlhäuser.

As a result, two major architectural challenges remain: interaction and integration. This raises the question of architecture. How will we design systems that facilitate integration? And how can we do so in a way that optimizes not only the technology, but also the human interaction? As Prof. Mühlhäuser said, "There's no architecture discussion about the integration of all these things without an interaction discussion about how we put the user in the center of all this."

There are some nascent efforts to meet this challenge. SAP recently teamed up with other partners on the Smart Products EU project. The project, funded by the European Commission's Seventh Framework Programme, includes ten partners from industry and academia. The initiative seeks to develop technology and techniques for building products with embedded "proactive knowledge." These smart products might carry information about themselves, their history, their users, and the environments and objects they have encountered. We will see many examples of these types of products throughout this book.

One way to move toward this vision is to give users "ubiquitous access," as in the model of the smart factory. Imagine a manufacturing plant where every worker has a personalized device capable of hands-free voice interaction, location awareness, processing, memory, federation with other devices, and trusted cooperation with services and things. These devices would be tickets to a full-fledged smart environment.

Unfortunately, added Prof. Mühlhäuser, this environment remains elusive. This vision requires a vast amount of technical work and many layers of support. "We really have six years of tough work to go if we really want to reach that in 2015," he said.

## Privacy, Security, and Trust

Moreover, the solutions must be more than technically sound; they must be trustworthy. "Indeed, I would claim the issue of privacy, security, and trust is about as underestimated as nuclear power safety in the '80s," said Prof. Mühlhäuser. "We need dialogue with the customers and with the public to lower the barriers of threat and the emotional attitudes of people. We have to take this seriously."

Unfortunately, the technical solutions still are not up to the challenge. "Believe me," added Prof. Mühlhäuser, "these problems are really not solved, not even in terms of security, let alone in terms of privacy."

*"Indeed, I would claim the issue of privacy, security, and trust is about as underestimated as nuclear power safety in the '80s."*

*— Prof. Mühlhäuser*

But the barriers aren't just technical. Privacy, security, and trust remain huge barriers for the average consumer. We need to make significant advances in privacy, security, and trust and engage in a public dialog. "Privacy and security are not only technical issues," he said. "They're also emotional issues that we have to address."

For the IOT, this is a live-or-die issue. Prof. Mühlhäuser said that future solutions will be "trusted or dead."

## New Architectures

The final challenge is scalability. How can the architecture of the IOT keep pace with the explosive growth? "From the architecture viewpoint, maybe the number one crucial point is scalability," said Prof. Mühlhäuser.

The number of nodes on the Internet is growing exponentially. CPU power has trouble keeping pace. Similarly, wireless bandwidth and battery life may also limit the expansion of the IOT. "It becomes more and more efficient to do things in a distributed fashion," said Prof. Mühlhäuser. "As this gap widens, distributed computing becomes a more viable alternative."

Architecture needs a new paradigm. In order to scale up, solutions will have to be long-living, open, and capable of self-X-capabilities like self-repair, self-configuration, self-healing, and self-discovery. Hierarchies will remain—and evolve a great deal. "Maybe not the hierarchies we have today," said Prof. Mühlhäuser, "but hierarchies are an excellent means for taming scalability problems and they'll remain for that purpose."

How do we begin to solve these problems? How do we begin to create the infrastructure that allows the IOT to move from dream to reality?

"We need to go ahead, envision the future, simulate it, do modeling—and not just go there and then see the disaster and then think of repair," Prof. Mühlhäuser said. "We have to think in advance and really tackle these issues."

## The Challenge

Dr. Rainer Zimmermann, Head of the Unit for Future Networks of the Directorate General for Information Society and Media at the European Commission, began his challenge with a provocative statement: the community of networking people remains ignorant of many of the concepts behind the IOT.

This lack of understanding is of particular concern to Dr. Zimmermann; he studies the network of the future for the European Commission.

According to Dr. Zimmermann, the networked society of the future depends on several pillars:

- An Internet that accommodates all users

- An Internet of contents and knowledge

- An Internet of services

These pillars must rest on a foundation of infrastructure with many elements. They include scalable and dynamic routing, adaptability, security, privacy, and trust, availability and ubiquity, and sustainability.

But this underlying architecture is a jumble of confusion, said Dr. Zimmermann. We should move away from hierarchical systems with their multiple layers of application services clouds, mediation services clouds, and connectivity services clouds. Too many protocols and types of services exist in this chaotic environment.

**Dr. Rainer Zimmermann** graduated in Engineering from the Technical University (TU) of Berlin in 1977 and obtained his Dr-Ing in 1985 from the same university. He worked from 1977 to 1985 as a researcher for the TU Berlin and the Fraunhofer Society. He then joined the Commission as a project officer for projects in the fields of production engineering, high-performance computing (1990), and software (1992). In 1995 he became Head of the Unit for Telematics between administrations.

After heading the Software and Systems unit (1999) and the Unit for Nanoelectronics and Photonics (2005), he currently heads the Unit for Future Networks of the Directorate General for Information Society and Media at the European Commission.

"This is a big mess today," said Dr. Zimmermann. "If you look at it, you will see many, many protocols supporting the application."

Meanwhile, several other forces are swirling into this mess, such as technological advances in microelectronics and nano electronics, as predicted by Moore's Law (which states that the number of transistors on a chip will double about every two years). Progress in photonics—an often-forgotten but essential element—represents another major trend. "The capabilities go up," said Dr. Zimmermann, "but the demands also are going up."

How do we get ourselves out of that corner? "Future networks should be structured by their relations," said Dr. Zimmermann. "They should not be highly structured, but incorporate some hierarchy. We want to decouple technical functions and physical functions from logical functions, from thinking."

## New Relationships

How do we define these relations? Dr. Zimmermann suggested a few of the major steps. We should start with the users. He defined these consumers broadly; they might include citizens, enterprises, and even devices such as sensors and actuators. Some of these are so-called "prosumers"—those who both produce and consume data. Other elements are content, services, and networks.

All of these elements are changing. Users, enterprises, and devices are evolving. Networks constantly evolve. Today we have broadband, wireless, and optical. In a few years we may have networks of networks, according to their applications. Content is changing and we have more peer-to-peer communication, new search media, entertainment, advertising, gaming, and so on. Finally, services are changing. Bring all that together and you have the IOT.

Traditional hierarchical relationships don't work because of the dynamic nature of these elements. For example, content is changing and we are witnessing a surge in peer-to-peer communication, new search media, entertainment, advertising, and gaming. Services are evolving from classical to smart services. In short, the targets are moving too much for rigid definitions and hierarchies.

Instead, these elements must be held together by their relations. For example, Dr. Zimmermann suggested that the relationship between users and networks should be determined by attributes such as naming, scalability, mobility, openness, interoperability, heterogeneity, and management. Similarly, the relationships between users and devices should be governed by attributes such as security, trust, service discovery, or context awareness.

Many questions remain unanswered. Is our vision of the future network the right one? Can corporate culture confront such a complex problem, or will it kill new ideas? Where are the entrepreneurs who will come up with new ideas and guide us into this *terra incognita*?

According to Dr. Zimmermann, the service delivery platform appears to be a promising way to bring together services, content, network, and users. But this just raises more questions in his mind. "Will the killer ideas come from industry, academia, or the public sector?" asked Dr. Zimmermann. "Initially I thought the answer was very clear: it will come from industry. But then I showed this slide to several people and many said, 'No, it will come from academia; industry is hopeless.'"

With so many open questions, where should we begin? Dr. Zimmermann identified several key issues:

- **Openness:** Is the industrial structure open enough to allow innovation?

- **Generalized platforms:** In Europe, the idea of shared platforms is popular. Everybody buys into the concepts of open platforms, interoperability, and standardization. "The problem is that it is not a business case," Dr. Zimmermann said. "Everybody wants them. Everybody wants them to be done, but by others, not by themselves, because the cost to build an open platform can be very high and the return does not necessarily come back to the one who builds them. So cooperation between companies is key."

- **Standardization:** What should the standardization process be? In recent years, standardization has increasingly been influenced by intellectual property law—which often stifles innovation. For example, the Chinese competitors in radio claim they pay 20 euros per device just to get into the game of 3G technology. "That is a perversion of that process," said Dr. Zimmermann.

- **Regulation:** Will it help or hinder progress?

- **Risks:** Will users accept the risks associated with new technologies that are seen by some as intrusive?

Dr. Zimmermann concluded by repeating the same refrain from the beginning: we must always bear in mind the perspective of the user.

## The Discussion

Once again, many questions were thrown out.

How should we approach the question of infrastructure? Should we worry less about infrastructure and more about applications? What should the role of government be? What about standards? And what will infrastructure look like?

The question of infrastructure of the IOT is not an easy matter—even for the insightful and well-informed participants at the IRF. "The question we had is whether the infrastructure for the Internet of Things will be rather different from the infrastructure we have in the classical Internet," said Prof. Dr. Friedemann Mattern of the Department of Computer Science and Institute for Pervasive Computing at ETH Zurich. "This is an important question, but there's no good answer. Is it something that you just have to improve? Or is it something radically new?"

How can we build the infrastructure of the IOT when we don't yet understand its applications? As Prof. Mattern notes, these applications are likely to be very different from those of the classical Internet. Standards represent another conundrum. The Internet rests on a bedrock of successful standards such as IP, HTTP, and XML, but it remains unclear whether the IOT will arrive at its own basic protocols.

"It will be difficult to have a few well-established standards," said Prof. Mattern. "So probably, there will be many standards, a hierarchy of standards."

Prof. Günther from Humboldt University suggested we should think about infrastructure not solely in terms of hardware. Rather, we should think in multidisciplinary terms.

*"If you only think about technical infrastructure, you're missing out on the economic and social aspects."*
*— Prof. Günther*

"If you only think about technical infrastructure, you're missing out on the economic and social aspects," he said. "The people aspects are just as important to make things succeed."

We need use cases, Prof. Günther said, but use cases are tied to existing models. Maybe we need to think more radically. Perhaps the IOT could bring new breakthroughs and killer applications in areas like energy conservation and healthcare.

"There's a certain danger in seeing the Internet of Things as evolutionary rather than revolutionary," said Prof. Günther. "Maybe we're missing some of the great potential of the Internet of Things by not thinking enough out of the box."

Others had more ideas going in the same direction.

Prof. Dr. Daniel Kofman of Telecom ParisTech University in Paris and CTO of RAD Data Communications foresaw a complex infrastructure.

"The network of the future, the Internet of the future, the Internet of Things will be polymorphic," he said. "We will have interworking of very heterogeneous systems."

Standardization is the key issue. In the future, clouds of different technologies will interact. "The routing won't be based on addresses," added Prof. Kofman. "It will be based on identifiers—identifiers of services, identifiers of people, identifiers of things. The routing will be semantic."

The routers will not be routers in the traditional sense. "They will be gateways that will have the intelligence to understand the services and choose the right path at the service level based on this," Dr. Kofman continued. "And this will be dynamically composed because, per service, you will need these gateways to dynamically adapt, to integrate a new functionality."

*"The routing won't be based on addresses, it will be based on identifiers— identifiers of services, identifiers of people, identifiers of things. The routing will be semantic."*

*— Prof. Kofman*

This future scenario raises key questions: how can we standardize identification management? And how can we standardize the semantics for components to describe themselves in order to be composed?

Prof. Khosla of Carnegie Mellon predicted that the IOT will be heterogeneous. "There's going to be no single architecture that's going to be good for different types of applications," he said.

What will these architectures then look like? Speaking on behalf of his group, Prof. Khosla suggested three common properties.

There has to be some commonality at some bottom layer (for lack of a better term, they suggested the description "bit pipe").

Prof. Khosla's group foresaw three data varieties. The first is characterized by small packet size, high data rate, real time, and low latency. The second is more traditional, with moderate packet sizes and data rates. The third is streaming data. "We would have to have infrastructure that supports all three types of data rates and it has to be integrated," he said.

The systems above this pipe level have to be differentiated—thus different architectures and different applications. "Healthcare applications have different architectures than car networking or data networking," Prof. Khosla said.

In the middle, there has to be some notion of virtualization that brings all of these elements together and allows composition of different applications.

This brought Prof. Khosla to another perennial question: what should the role of government be?

"There has to be infrastructure that's invested in by the government," said Prof. Khosla. "Then there are other layers of infrastructure that are application specific, user specific, and that has to be developed on a case-by-case basis."

Other questions followed. Will the IOT be driven by infrastructure or applications? Indeed, the IOT will mean little without applications. Forum moderator Dan Woods played devil's advocate.

"Should we just declare the Internet of Things open for business, wait for a good application and forget about the infrastructure?" asked Mr. Woods.

This drew a quick response from Dr. David Skellern from NICTA, Australia's Information and Communications Technology (ICT) Centre of Excellence.

"That won't work at all," he insisted. "We'll just end up with all of these disparately connected devices that actually don't have any use to anyone except the person who put them in place. So we absolutely have to have an application."

We also need models and incentives.

"We need to be creative around business models and financing hurdles," said Claudia Funke, Director of the Munich office of McKinsey & Company. Too often, she added, businesses are "dis-incentivized" to innovate. "In that area, I do see a certain role for governments, what I would call smart regulation and smart incentives."

Prof. Dr. Karl Aberer, Ecole Polytechnique Fédérale de Lausanne in Switzerland, brought up the failed burglar alarm. Why do these early IOT applications so often fall short?

"Business data is very different from Internet of Things data," he said. "It's discrete and not continuous. It's about logical relationships and not about statistical relationships. And I think a major reason for application failure is that the whole infrastructure, from the networking up to the application systems, really does not know how to represent, how to interpret, and how to deal with this new type of data in the Internet of Things."

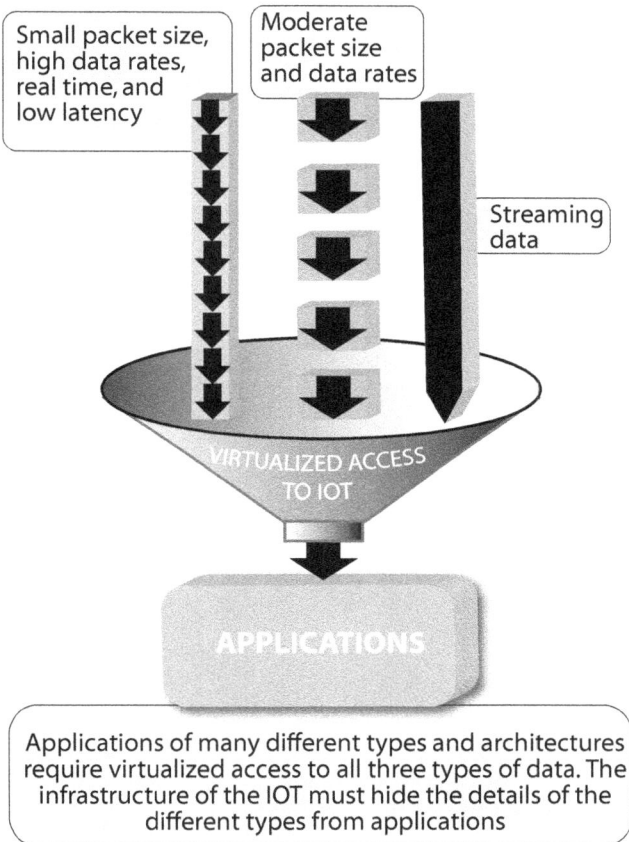

*Figure 3-2. Virtualized Access to the IOT*

4

What makes a killer application in the Internet of Things? According to Prof. Dr. Hao Min, Research Director of Auto-ID Labs and Professor of ASIC & Systems State Key Laboratory at Fudan University, a discussion of killer apps should begin with a simple question: who benefits?

## The Vision

Prof. Min returned to the often-cited example of the Wal-Mart RFID system. Why can't the Wal-Mart system meet the mass adoption milestones? Prof. Min recalled two conversations at a recent industry conference that highlighted these shortcomings. He asked one Wal-Mart executive why the retail giant mandated mass adoption of tags. The Wal-Mart exec replied

**Prof. Dr. Hao Min** is Research Director of Auto-ID Labs and Professor of ASIC & Systems State Key Laboratory at Fudan University. He is also the Chairman and co-founder of Shanghai Quanray Electronics, which he started in 2006.

Prof. Min got his Ph.D. from Fudan University in 1991 and then worked in the ASIC & Systems State Key Laboratory. From 1995 to 1998, he worked as a visiting associate professor in the department of electrical engineering at Stanford University.

From 1998, he served as Professor and Director of ASIC & Systems State Key Laboratory, Fudan University and worked on smartcard and RFID chip technology. He started the Auto-ID Center China in 2002, serving as Research Director. He also served as General Manager of Shanghai Huahong IC Co. Ltd. from 1998 to 2006.

Prof. Min's research areas include VLSI architecture, RF and mixed signal IC design, digital signal processing, and image processing. He has published more than 50 papers in journals and conferences. He is the inventor of more than 10 patents (pending).

that the company wanted to build the infrastructure in its distribution center and retail stores and facilitate mass adoption.

Then Prof. Min talked to a Wal-Mart vendor who told a different story about the RFID system. Initially, this company was excited about tagging and was eager to be among the first 300 suppliers who adopted it. After all, Wal-Mart touted the benefits for the vendor, such as the ability to track products and harvest all kinds of data. In practice, however, the arrangement proved not so attractive to the suppliers. The vendor bears much of the cost for this RFID system and tagging the pallets. This vendor participated in the pilot run and expected to have the RFID information at their fingertips. Not so fast. Wal-Mart said their IT infrastructure could not support that kind of information sharing right now. As Prof. Min concluded, "The vendor pays the cost, but right now they cannot get the benefit that they expected."

This brought Prof. Min back to killer apps. Ideally, a killer app should benefit all parties. Otherwise it isn't really a killer app. Wal-Mart's RFID model, he suggested, is a failure because it doesn't benefit all parties.

*Ideally, a killer app should benefit all parties. Otherwise it isn't really a killer app.*

But there are more promising examples—like the Chinese equivalent of Wal-Mart. Prof. Min's company works with the Brilliance Group, one of China's largest retailers. Based on what he has seen so far in Asia, Prof. Min contended that "secured supply chain management" is a potential killer app for the IOT.

## Linking the Supply Chain

Consider the typical supply chain. It runs from factory to wholesaler to distribution center to retailer and, finally, to the end consumer. There are many steps along the way...and many opportunities to violate the integrity of the process and product. Goods can go astray or be relabeled. Products can be stolen, swapped, or counterfeited.

"The final end users, because of counterfeiting and the like, sometimes do not really know what they are getting," Prof. Min said.

For example, in China, tobacco products vary between provinces. "Because of the price difference there are some guys who just switch," said Prof. Min. "They've got the low cost from the manufacturer and they sell that in a high-price province. The end users get a counterfeit, something that is completely fake."

Sometimes products are disguised with bogus labels. This may include tobacco products or, more disturbingly, pharmaceuticals. Sometimes drugs are relabeled with incorrect doses to increase profit margins. One such scandal in China resulted in the death of several people, spurring the government to take action with a better tracking method.

How to prevent such abuse? A secure supply chain needs several elements, according to Prof. Min. It needs serialization (unique identification for each product), authentication (ability to confirm product identification), integrity (prevents unauthorized changes in information), non-repudiation (sender cannot deny), and traceability (the ability to track product history).

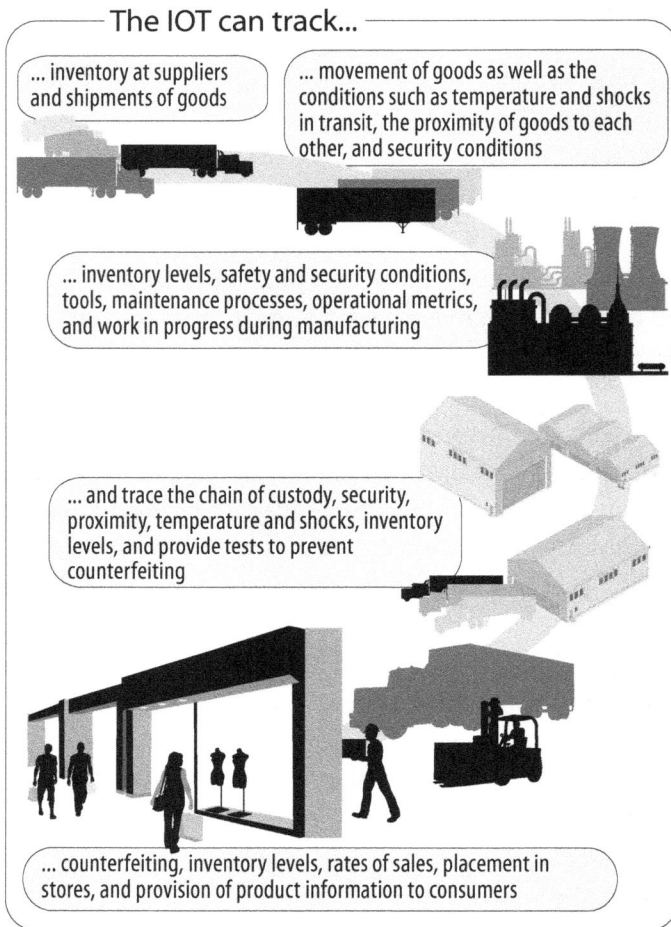

*Figure 4-1. Linking the Supply Chain*

In this case, they turned to RFID. RFID is difficult to clone, rewritable, performs well, doesn't need line of sight transmission, and is affordable. These tags are capable of storing up to 1k of memory and, added Prof. Min, capable of tracking their own movements—a key difference that distinguishes them from the Wal-Mart model. In contrast, Wal-Mart used tags with only serial numbers or EPC codes. Thus far the system has performed well, although it remains cost-effective only at the box level, not for individual items.

Prof. Min expanded on this vision and outlined how such an RFID system might change the supply chain. At the factory, the product is packaged with a chip that holds information about the product identity. Each product has a unique digital signature. It transmits information via radio waves to a reader. Information flows back and forth between the database and the various steps in the process, from factory to consumer.

According to Prof. Min, the Shanghai Food and Drug Administration established one such system. Companies must report to the FDA at key points on the supply chain: factory, wholesaler, distributing center, and retail. The FDA knows exactly which product is in which location. In the event of a recall, they can accurately locate all the drugs and take them off the shelves.

This system stores the product ID number, digital signature, and other information. The reader checks the digital signature to verify whether it came from the original manufacturer, thus reducing the odds of counterfeiting.

A manufacturer puts an RFID tag on the box and sends the info to the FDA. The distribution center confirms when it arrives and when it leaves and once again sends the information to the regulatory agency. This process continues throughout the length of the supply chain. Authenticity can even be checked by hospitals, retailers, or consumers. At the retail level, there are kiosks where buyers can scan the packages and determine whether they came from the original manufacturer.

This model, said Prof. Min, offers a glimpse of how the IOT may transform supply chain management. Those tiny RFID chips may be the killer app for the IOT.

## The Challenge

The next speaker began with an applause line: he promised he wasn't going to talk about RFID.

Justin Rattner, Vice President and CTO of Intel, has worked on the IOT for at least a decade and thus has the unique authority to describe its shortcomings. He summed it up with a pithy bit of caution: "The more we learn, the less we know."

*"The more we learn, the less we know."*
*— Mr. Rattner*

This is not simply a technological problem. Technologists have come up with plenty

**Justin Rattner** is Vice President and CTO of Intel. He is also an Intel senior fellow and head of the Corporate Technology Group. In 1989, Mr. Rattner was named Scientist of the Year by *R&D Magazine*. In December 1996, he was featured as Person of the Week by ABC World News for his visionary work on the Department of Energy's ASCI Red System. In 1997, Mr. Rattner was honored as one of the 200 individuals having the greatest impact on the US computer industry. He has received two Intel Achievement Awards for his work in high-performance computing and advanced cluster communication architecture. He is a member of the executive committee of Intel's Research Council and serves as the Intel executive sponsor for Cornell University. Mr. Rattner is also a trustee of the Anita Borg Institute for Women and Technology. Prior to joining Intel, he held positions with Hewlett-Packard and Xerox.

of solutions that were more than adequate to address the problems at hand, yet these technologies still haven't turned into the killer app. Why? Intel's work on wireless sensor networks highlights some of the barriers and serves as a cautionary tale.

According to Mr. Rattner, Intel developed a system of ad hoc wireless networking with very small nodes, ample non-volatile memory, and low power requirements. They were sure they had a winner.

"After all of that work over the decade, we still found ourselves looking for the killer application," recalled Mr. Rattner. "We had great technology. We were sure that we'd just take that technology and, man, you know, we'd just be knocking down one opportunity after the next."

It didn't work out that way. They struggled to find practical applications for this technology. They leveraged Intel capital, invested in startups, partnered with entrepreneurs, big companies, and small companies and did everything they could think of to turn the technology into a successful application. The research team even hired some business types to find a killer application.

"I tolerated a year of this while they went out and looked high and low for the big win that would use the technology," recalls Mr. Rattner. "I finally had to pull the plug, because, after a year, they were still struggling to figure out a business model...At the end, we essentially pulled the plug."

Intel continued to develop this technology and still uses sensor and wireless networks in other business lines, such as health products. But these applications are far more modest than the grand visions that once danced in their heads. As Mr. Rattner said, "This notion we had of building a business around wireless sensor networks really has failed to materialize."

The lesson: it's not just about the hardware.

"Technology is not the problem, as far as we can tell," said Mr. Rattner. "It can certainly help solve specific problems...but it's not the thing that prevented us from finding a successful business opportunity here."

## New Models Needed

So what was missing?

In order to fly, these technologies need business models. "We assumed that a horizontal business model would emerge in the area of wireless sensor networks," said Mr. Rattner. "The idea was, well, there'll be hardware suppliers and software suppliers, and the operating system suppliers, database suppliers—sort of the standard computer industry model that emerged during the PC era."

In the end, Intel passed on the RFID business because it was not compatible with its main semiconductor business. The company specializes in powerful chips with high intellectual content, not those puny little RFID chips with small memory and virtually no intellectual content. The company also was disadvantaged by its starting position. The horizontal model didn't work—and even now remains a long way off.

"It might happen," said Mr. Rattner. "There may be some kind of a black swan in all of this, but right now, we seem to be some distance from a horizontal model."

In fact, the companies that have been successful in the field of sensing tend to employ vertical models. For example, applications within factory settings tend to be narrow and specialized.

Another cautionary tale comes from a pilot project that Intel pursued with BP, the multinational petroleum company. Intel deployed a sensor network for a tanker that shuttled to the oil fields off Scotland. "BP loved this—they absolutely loved it, and they would have bought it for all of their ships and deployed it," said Mr. Rattner. "But we couldn't find anyone who actually wanted to focus on the vertical application of deploying these sensor networks onboard ship."

Again, it wasn't about technology. You can talk about technology to a customer like BP until you're blue in the face. "In the end," added Mr. Rattner, "it's whether you can field the solution or not."

### Consumers in the Driver Seat

If not technology, what will drive the IOT?

"I'll leave you with a bit of a radical thought," Mr. Rattner said. "I think at this point that the Internet of Things is actually going to be driven by consumers."

We already have seen many examples of technology that first took off in the consumer space and later infiltrated the enterprise. Social networking is a prime example. Energy efficiency might be the next example; right now consumers are driving interest in new technologies and conservation. Similarly, Mr. Rattner suggested, consumer demand may drive interest in the IOT.

Another driver might be technology that makes more energy efficient buildings. If so, "sensorless" sensing and perception may hold the key to the next killer app. The technology would find a match with the right model. At the moment, however, the fate of sensing technology is unclear. It remains in what Mr. Rattner described as the chaos phase.

*"The more that we can find ways to sense things with existing technology, the better off we'll be."*
*— Mr. Rattner*

"There are Darwinian processes at work here and they're fundamentally important," he said. "If you get into over-specifying things and bringing all the experts together to figure out what to do, you prematurely reign in that chaos...There's something to be said for letting the process of natural selection occur."

Perhaps a better approach is to wait until a more highly evolved model crawls out of this primordial ooze. Only then, added Mr. Rattner, should companies move to define the infrastructure and set standards.

"The more that we can find ways to sense things with existing technology, the better off we'll be," he said. "Who would have thought of the huge number of applications that would have emerged from the iPhone as a result of having the accelerometer built? Every day there are wild and crazy applications that emerge for the iPhone. So using existing sensors may be key to driving the Internet of Things."

## The Discussion

What kind of killer applications will emerge from the IOT? The vision and challenge speakers sparked a flood of ideas in the group discussions. The group floated many ideas for potential killer apps: energy management, healthcare, physical exercises, tourism, real-world gaming, transportation, crisis management, and environmental monitoring.

What are the requirements of a killer application?

Dr. Stan Smits, Senior Vice President and Chief Software Technology officer of Philips Healthcare, offered a few suggestions. It needed mass adoption, ease of use, and an attractive user interface. Killer applications are not made only by the application *per se;* they also depend on the overall platform, ecosystem, and the business model. A prime example is the Apple App Store.

Others built on this idea. "That platform has to have more characteristics than just the functionality," said Ms. Murphy-Hoye from Intel. "It has to be a platform with features that also drive accelerated mass adoption."

Others suggested other characteristics of killer apps, such as user experience, reliability, usability, public acceptance, and clear benefits.

One potential killer app that emerged repeatedly was energy management and sustainability. Perhaps sensors could help create smart buildings capable of regulating their own power consumption, thus providing important tools for sustainability and reducing climate change.

Claudia Funke of McKinsey & Company championed these green killer apps. She suggested applications for carbon tracking and energy-efficient

| Examples of Killer Apps and Their Functions | |
|---|---|
| **Killer App** | **Function** |
| Energy management and sustainability<br><br>Carbon tracking and energy-efficient production<br><br>Environmental monitoring | Smart buildings capable of regulating their own power consumption<br><br>Facilitating with regulations for carbon transparency and tax incentives |
| Smart homes | Smart homes that would allow the elderly to continue living independently for longer periods |
| Transportation and logistics | Seamless and consistent management and tracking of product history between all security-related organizations |
| Crisis management | Platforms facilitating the simulation of disasters and the collaboration between all security-related organizations |
| Urban management | Platforms supporting municipal administrators with real-time interpretation of data from all kinds of sources based on sensors and state-of-the-art communication technology |

production. Government would not necessarily dream up these models, she said, but it could play a vital role in facilitating them with regulations for carbon transparency and tax incentives.

Another popular suggestion for killer applications was healthcare, particularly smart homes that would allow the elderly to continue living independently for longer periods. Apparently the failed experiments described by Prof. Österle, University of St. Gallen, did not deter advocates who saw healthcare as a fertile field for the IOT.

Yet, Prof. Heuser of SAP quickly challenged the dream of killer applications in healthcare. The players in the market simply make it impossible, he said.

"I think there is no business case for healthcare," Prof. Heuser said. "Not because there is no reason for it, but because the players out there do not really allow it. I've never seen a killer app working in the healthcare business."

Dr. Kaiserswerth of IBM added more caution on health applications. It's hard enough for an able-minded person to pair two Bluetooth devices, he said. How can we expect elderly people, especially those with dementia, to troubleshoot a smart home packed with devices. Even so, others remained undeterred. Dr. David Skellern, NICTA, acknowledged the challenges, but added, "If you can get around those obstacles, these are killer apps."

Prof. Dr. Janet Wesson, Nelson Mandela Metropolitan University (NMU) in Port Elizabeth, South Africa, also continued to tout the healthcare domain as a ripe area for killer applications. "We think there is a successful use case out there already, and that use case is well established." she said. "It's a pacemaker."

5

How will the Internet of Things transform the factory? It already has begun to do so, but mostly remains confined to internal applications within the firewall. Even so, the factory is prime territory for the IOT. Factories are closed environments, where business owners can see quick return on their investments and tangible results in visibility, added value, efficiency, and new services. In the future, smart devices will increasingly spread into the factory and automate more and more processes. But it remains to be seen how quickly these processes will spread outside the factory walls into more open loops.

## The Vision

According to Rick Bullotta, co-founder and Chief Strategy Officer of Burning Sky Software, the IOT remains largely the *Intranet* of Things.

**Rick Bullotta** is the co-founder and Chief Strategy Officer of Burning Sky Software, a pioneer in collaborative, real-world-aware applications. Mr. Bullotta was previously CTO at Invensys Wonderware, a leading global provider of manufacturing operations software solutions, and was a vice president with SAP Research.

Mr. Bullotta was also the co-founder and CTO of Lighthammer Software Development, where he was responsible for conceptualization and development of innovative web-based and service-enabled software products targeted at the manufacturing industry. At Lighthammer, Bullotta identified and created a new market segment for "manufacturing intelligence and integration" software. Lighthammer was later acquired by SAP. He has contributed to a number of industry standards organizations including ISA-95/IEC 62264 and the OPC foundation.

Mr. Bullotta has also been involved in the industrial sector in diverse roles, including factory operations and management, systems integration, sales and marketing, product management, and industrial engineering. He holds a degree in operations research and industrial engineering from Cornell University.

The technologies, like those of the Future Factory, have been deployed mostly in closed environments like manufacturing plants.

How can this vision expand into a full-fledged IOT? According to Mr. Bullotta, the IOT must incorporate three elements that he labels "meatspace" (people), "cyberspace" (systems), and "atomspace" (things). All of these elements must be woven into a network of collaboration.

*"It's not that we're just going to be communicating between sensors and people, or sensors and systems, or systems and people. It's all of the above."*
*— Mr. Bullotta*

"It's not that we're just going to be communicating between sensors and people, or sensors and systems, or systems and people," said Mr. Bullotta. "It's all of the above."

The Internet still has much to learn from the industrial world, and vice versa. The context is not just the factory, but the larger world in which

To provide value in manufacturing, the IOT must bring together three dimensions into a collaborative environment

**MEATSPACE**
The people in the plant must be able to communicate with each other and also understand the state of the atomspace and the systems

**ATOMSPACE**
The IOT must report on the state, position, temperature, location, and other information about the things in a plant

**Cyberspace**
The IOT must provide information to the systems in the plant which will be used to gather and display information to support collaboration

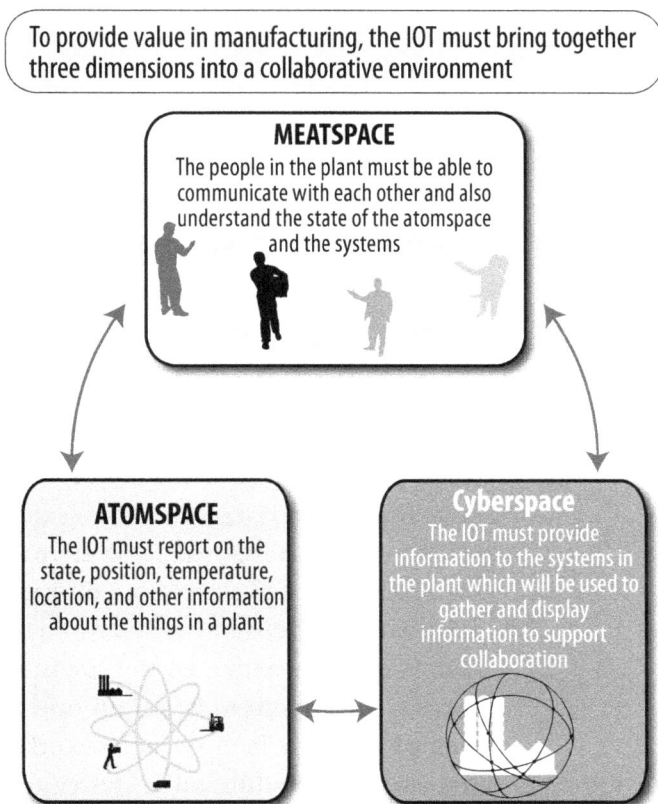

*Figure 5-1. The Three Elements of the IOT*

that factory operates. "You can't just look inside the four walls of a manufacturing plant to think about a manufacturing plant," said Mr. Bullotta. "It's a much, much bigger ecosystem." (See Figure 4-1 on page 40.)

How to transform this ecosystem into a smarter environment?

Mr. Bullotta suggested a few key steps:

- Applications must be event-driven

- They must be instantly adaptable to the chaotic conditions on the factory floor

- They must be suitable for quick reconfiguration and new work processes

Despite all the emphasis on machines and technologies—the things of the IOT—we never should forget about the role of people. The systems must be adapted to humans. They must take into account humans limitations and the very un-machinelike nature of human intelligence.

"I used to run heat-treating operations in a steel facility at the start of my career," said Mr. Bullotta. "You learn pretty quickly that humans are one of the major sensors and actuators in manufacturing processes."

We must find new ways for manufacturers to operate, collaborate, and innovate with technology. According to Mr. Bullotta, part of the solution is to arm workers with more information and devices to help them collaborate more effectively. "I'm a big proponent that a shop floor worker is an information worker, but they're not being treated as such," he said. "They need to get a bit more respect, quite frankly, and be given the tools they need to leverage their brainpower to be more effective."

*"I'm a big proponent that a shop floor worker is an information worker, but they're not being treated as such."*
*— Mr. Bullotta*

Finally, we should explore multiple modes for interaction between humans and machines. People can't be expected to type in text messages—the keyboard is not an option on the floor of a chaotic factory, where workers might be wearing gloves or cumbersome coveralls while shuttling between five machines.

How will the IOT seep into the manufacturing world? Mr. Bullotta foresees a process of "progressive automation." There is a spectrum of increasingly complex tasks that goes from observation to decision to action. Machines will gradually perform more and more tasks along this spectrum.

"There's an evolutionary process we'll see in the adoption of technologies around the Internet or intranet of things," said Mr. Bullotta.

In the first generation of the IOT, machines are likely to be cast as sensors. They will be observers that bring real-world awareness, but humans will remain the principal decision makers. Over time, technologies like artificial intelligence and assisted decision making will enable machines to shoulder more responsibilities. In some cases, Mr. Bullotta said, machines may perform all three roles and completely "close the loop."

## Manufacturing Deconstructed

The IOT may help to hold together an increasingly fragmented value chain.

A homogeneous manufacturing company is becoming a thing of the past. Few companies provide all their own services along the entire length of the value chain. More often, the process is broken down and pieces are entrusted to partners. For example, one partner might handle product design, another does manufacturing, and still others handle marketing, brand ownership, distribution, servicing, and so on.

"You look at the end-to-end lifecycle of a product, it quite often touches multiple businesses and constituencies," said Mr. Bullotta. "Yet the information that comes out of each of those activities is very valuable to all the other functions in the organization. So, we need to explore collectively, how we can create a technical infrastructure that allows that information to be shared across the value chain?"

Of course, this requires management of information and selective access across the firewall. An analogy is an airliner: it has its own internal systems for communication and navigation, and it also has systems that connect it to the larger air traffic control system.

We need ways to make information flow through the silos of the intranets. The upside is very attractive: people throughout the value chain can benefit from data generated at other points in the value chain—information that they probably would not see otherwise. Imagine an enterprise that makes washing machine motors. A manufacturing executive would benefit from visibility into product design by getting information about the machines that the factory will soon be making. Similarly, a service person would benefit from being able to see what's happening in the manufacturing side. The information can flow backward or forward. Say a motor has a 16% failure rate—this information would be immensely valuable to the manufacturers and help them make changes to ensure better reliability.

According to Mr. Bullotta, manufacturing stands to benefit from both intranets and the IOT. A cloud model (a topic discussed extensively in the IRF 2008) seems particularly fitting for these scenarios. Of course, this raises important questions. How do we build new firewalls that allow this kind of selective-yet-open access? How do we deal with issues like security and trust?

"Maybe that line starts to blur," said Mr. Bullotta. "Maybe the factories truly can become more opened up directly to the Internet if some of those issues are dealt with."

## Sustainability Implications on Manufacturing

Sustainability is a hot topic these days—and one area where the IOT can make a major impact. Mr. Bullotta foresees unique opportunities for companies like SAP to integrate sustainability and energy management into business planning. In fact, he has completed many similar projects for clients who wanted to build systems for managing energy use. For many companies, energy has become a key driver in planning and cost processes; if these companies had better visibility, they could plan more accurately, take advantage of pricing opportunities, and be more agile in seizing opportunities.

"Prices would be more dynamic," Mr. Bullotta said. "You're going to need to be able to adjust your prices, get in and out of markets when you have a competitive advantage because of energy prices."

These systems even can make product design more sustainable and environmentally-friendly. Mr. Bullotta suggested they may help us move away from disposable products with short life spans. A car, for example, could be seen as a platform that is not junked at the end of its lifespan but *upgraded* with new components, like mechanicals, décor, or electronics.

"We certainly have a disposable mentality in consumer electronics today," he said. "It would be interesting to see an application where we start to see these things more as platforms that are innately upgraded."

## Long Tail Opportunities in the Manufacturing Value Chain

If the vehicle becomes a platform with a longer life span, it also becomes an opportunity to sell new components and upgrades, such as infotainment, new seats, tires, or servicing. This approach allows companies to expand the duration and depth of their relationships with customers and opens the door for cradle-to-grave product lifecycle management. Mr. Bullotta pointed to the iPhone and Apple's App Store as an example of this long tail wagging happily and profitably.

"I don't see why that same relationship couldn't theoretically exist for many other kinds of durable goods," he said. "I think we'll see a trend in that direction as well."

Cloud computing—also discussed at length in the IRF 2008—represents another potential opportunity, and challenge. The cloud offers the promise of putting information in a central place where everybody throughout the value chain can have access to it. Sounds nice, but Mr. Bullotta, a software expert, knows this is easier said than done.

"Getting that information up to a common place is not a trivial problem, because it comes from many different lines of business applications," he said.

Uniting all these systems poses other challenges, such as semantic linking and trust. "We've got a lot of hard work to do with that," he said. "We've got to come up with creative technologies to overlay new ontologies and semantic linking to these existing systems."

## Impact of Advanced Technologies on Future Manufacturing

How will these new technologies change manufacturing?

Increasingly, machines will replace or supplement human decision-making. Artificial intelligence and adaptive decision-making will continue to play a role but these systems can only do so much. Mr. Bullotta echoed his earlier theme: manufacturing is chaotic. Systems are developed on the fly; future technologies must adapt to this improvisational way of working. The SoKNOS system (described in Prof. Heuser's introductory speech) is a prime example of this adaptability.

"Provide people with the information in systems to do guided decision support, and, over time, let's look at applying advanced computing technology, learning systems, and so on to take some of that burden off the human," Mr. Bullotta suggested. "Dealing with the ambiguous is something that people do reasonably well. That's not a trivial thing to do in a generic fashion in software."

Robotics will play an increasing role. Today robots are used for repeatable tasks like welding specific parts. These tasks don't require discretion or decision-making, just the ability to stick to a monotonous routine—an ideal task for a machine.

"But to see robotics move into different kinds of applications, tasks the humans do today, is going to require much richer and more adaptive intelligence," Mr. Bullotta said. "A true robot should be an autonomous thing."

Will such machines ever achieve the ideal of a lights out factory— one where machines do virtually all of the labor? This vision is like nirvana—something that we strive for but never quite achieve. "There's always going to be maintenance people keeping the equipment alive," said Mr. Bullotta. "They're going to need lights to do their job."

### Social, Economic, and Governmental Implications

The IOT also raises a host of questions for society at large.

According to Mr. Bullotta, we must be prepared to confront an array of vexing social questions. What happens when we start replacing workers with technology? How do we find a new place for these laborers? How do we train the next generation manufacturing worker to become more of an information worker?

Similarly, we are likely to face an array of economic questions. Will the long tail create more monopolies? How does it change the possibility for fraud, criminal behavior, or safety risks? Who bears responsibility for what? Who should customers contact for support? "Now that we've broken our value chains up across 10 companies, who do you blame if something goes wrong?" asked Mr. Bullotta. The brand owner often bears the onus of the fallout. This liability provides added incentive to improve visibility across the value chain with technologies like RFID.

Finally, we should think carefully about what this scenario might mean for government. How will regulations affect manufacturing competitiveness? In the US, there is widespread concern about the decline in manufacturing as a national security issue.

### Summary

Yet these concerns should not obscure the bottom line message. Despite all the uncertainties, Mr. Bullotta had no doubt that the IOT can bring huge gains in productivity and vast new possibilities.

"There is phenomenal opportunity looking forward," he said. "There's roughly four to five times as many factory workers as there are office workers. For those who can deliver solutions to those workers, it's a very, very significant opportunity."

## The Challenge

According to Prof. Dr. Wolfgang Wahlster of DFKI (the German Research Center for Artificial Intelligence) and Saarland University, the IOT will find fertile ground in future manufacturing. One area of particular potential is low-volume/high-mixture manufacturing—an area where Germany has done particularly well.

Prof. Wahlster turned his attention to a now-familiar topic: semantic product memory (also known as digital product memory) in the smart factory. In short, semantic product memory is a vision for smart products that carry their own life history in embedded systems. Consider the

**Prof. Dr. Wolfgang Wahlster** is the Director and CEO of DFKI, the German Research Center for Artificial Intelligence, and a professor of computer science at Saarland University. He has published more than 170 technical papers and 8 books on language technology and intelligent user interfaces.

Prof. Wahlster is a fellow at the Association for the Advancement of Artificial Intelligence (AAAI) and the European Coordinating Committee for Artificial Intelligence (ECCAI). In 2001 he was presented the Future Prize—Germany's highest scientific award presented by the President of Germany.

In 2002 Prof. Wahlster was elected full member of the German Academy of Sciences and Literature, Mainz, and was the first German computer scientist elected foreign member of the Royal Swedish Nobel Prize Academy of Sciences. In 2004 he was elected full member of the German Academy of Natural Scientists Leopoldina, and of acatech, the Council for Engineering Sciences at the Union of the German Academies of Science and Humanities. He serves on the Executive Board of the International Computer Science Institute (ICSI) at Berkeley and the EIT ICT Labs of the European Institute of Innovation and Technology.

pyramid model of factory automation. Smart product memories allow us to envision an integrated system from the devices on the factory floor to the human operators to the ERP systems in the executive suites. "From your business level, you can drive what is happening, in the sense of really high precision management of the factory, what is going on, on the shop floor," said Prof. Wahlster.

Smart factories are suited to service-oriented architecture. From bottom to top, everything is defined as a service—one that allows users to access all these elements with a portable device like a cellphone. This means we have to go from "bits and bytes" (electrical engineering), to functions (software engineering), to semantic services (semantic technologies).

This poses a formidable challenge of integration. For example, it must weave together the proprietary protocols of a tooling machine with

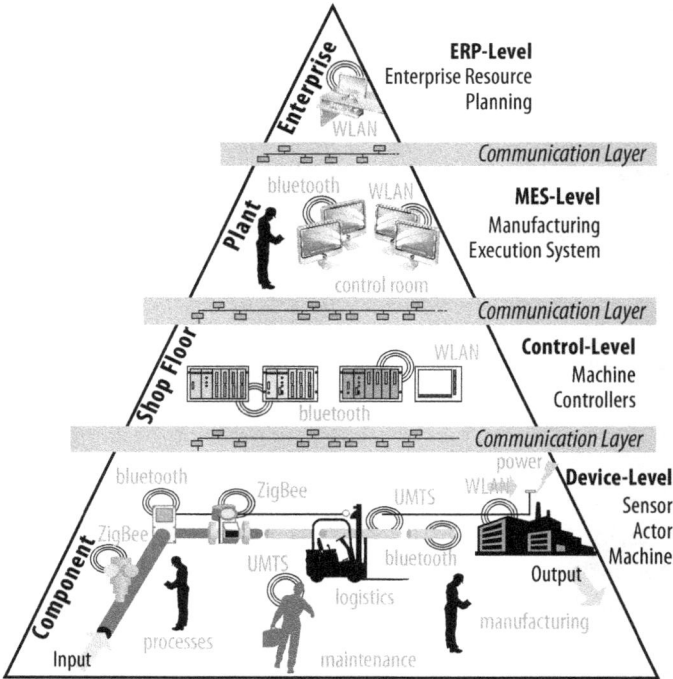

*Figure 5-2. The Pyramid Model of Factory Automation*

semantic ontologies. It must create a seamless whole from electrical engineering up to ontological engineering.

One model is the DFKI Living Lab Smart Factory *(http://www. smartfactory-kl.de/)* in Kaiserslautern, directed by Prof. Detlef Zühlke. According to Prof. Wahlster, this is the world's first multi-vendor research, development, and demonstration center for the factory of the future. It relies on the technologies that currently define the IOT: wireless, RFID, sensors, and service-oriented architecture. It serves as a living lab for interoperability testing of components from partners, including SAP Research, BASF, Siemens, Bayer, Bosch, and others.

The Smart Factory is completely wireless. This brings some challenges, such as an array of wireless protocols, but also many advantages. According to Prof. Wahlster, even blue-collar workers can access every piece of the factory via their cellphones. He predicts the smart phone will become the universal interaction device for future manufacturing. At the time of the IRF 2009, the facility was producing shampoo, and Prof. Wahlster joked that the only "wires" were the pipes that pumped the shampoo.

Such pioneering work inevitably encounters new challenges. According to Prof. Wahlster, one of the greatest hurdles is semantic product memory. These technologies essentially keep a diary of the entire life of the product. For example, the Siemens Future factory has used RFID antennas to read information from printed circuit boards for shipping MP3 players. Another example is blister packs of drugs, which can be personalized at the Kohn Pharma 7x4 factory with 400 different types of medication. These early use cases show how the IOT can add tremendous value. Prof. Wahlster contended that semantic product memory "is a killer application for the Internet of Things."

This semantic product memory, explored in Project SemProM *(http:// www.semprom.org)*, has the potential to revolutionize business. In logistics, it could help track the temperature of perishable foods as they are trucked to distribution centers. In retail, it helps monitor quality control and, for example, might alert a store when a product has been recalled. On the business side, it can help with production, logistics, distribution, retail, and quality control. On the consumer side, it also opens a new window of transparency and traceability and reassures customers about

Semantic product memory systems go beyond the basic functions of RFID
and other technologies oriented only to identifying products

### Input/Output Capability

Sensors to acquire
information about the
environment surrounding
the product

Means of communication
to broadcast and receive
information

### Information Processing Capability

To synthesize information and perform
computations

### Information Storage Capability

To retain the history of the product and related
information

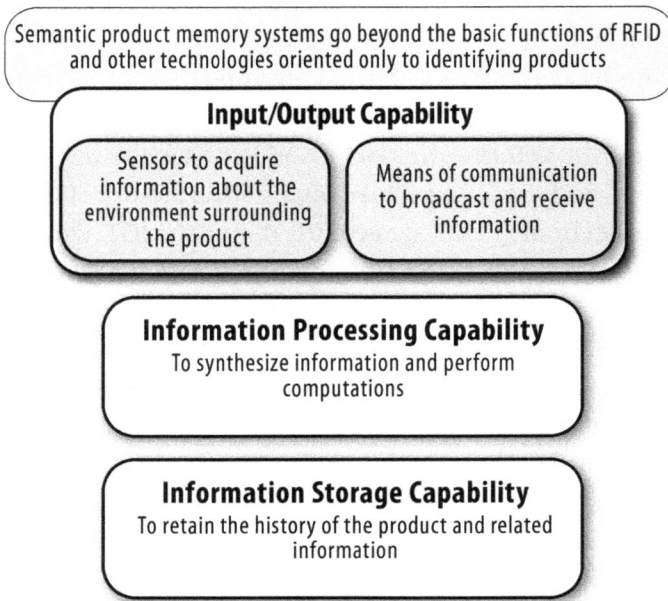

*Figure 5-3. The Basic Functions of Semantic Product Memory Systems*

the integrity of the product. For example, a shopper could check whether the tomatoes on a frozen pizza were grown organically.

"The pizza factories are very excited about it, because really, customers, at least in Germany, appreciate this type of biofood and are ready to pay quite a lot, about 30% more, if they actually can track," said Prof. Wahlster. For this premium, they can even interrogate their pizza. As Prof. Wahlster explained, "You ask the pizza, 'Where have you been? What was the temperature when?' You really have the full tracking record."

Semantic product memory even adds value long after the sale, said Prof. Wahlster. A buyer of a used BMW could use his or her cellphone to check the "black box" of that particular car—and discover the car's repair history and whether it has a third-party spare part that might be unsafe.

How to move toward this vision? Prof. Wahlster suggested that companies begin with existing technologies, like barcodes and first

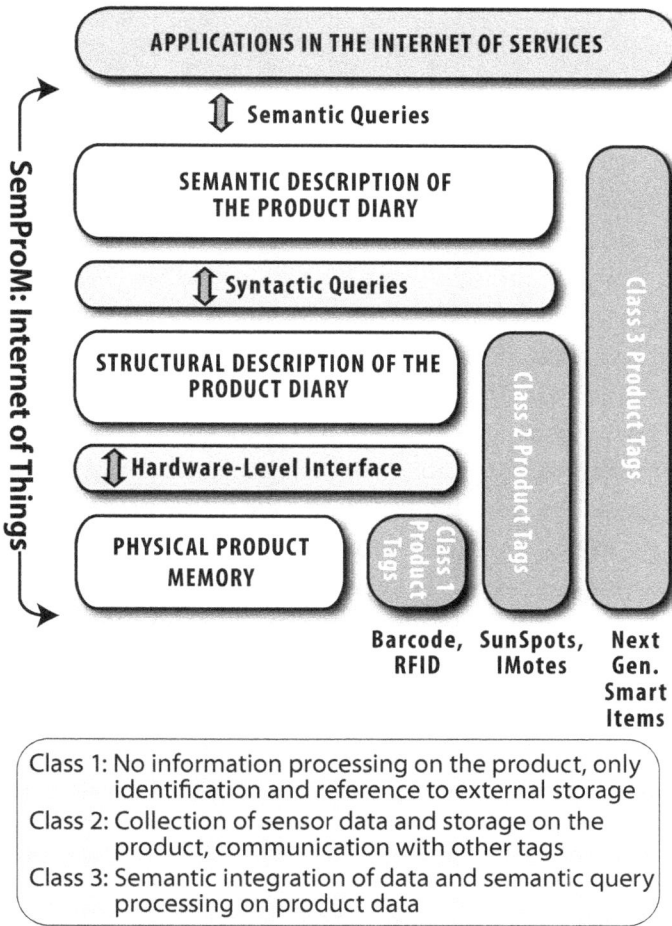

APPLICATIONS IN THE INTERNET OF SERVICES

⇕ Semantic Queries

SEMANTIC DESCRIPTION OF
THE PRODUCT DIARY

⇕ Syntactic Queries

STRUCTURAL DESCRIPTION OF THE
PRODUCT DIARY

⇕ Hardware-Level Interface

PHYSICAL PRODUCT
MEMORY

SemProM: Internet of Things

Class 1 Product Tags
Class 2 Product Tags
Class 3 Product Tags

Barcode,   SunSpots,   Next
RFID       IMotes       Gen.
                        Smart
                        Items

Class 1: No information processing on the product, only
         identification and reference to external storage
Class 2: Collection of sensor data and storage on the
         product, communication with other tags
Class 3: Semantic integration of data and semantic query
         processing on product data

*Figure 5-4. The Layered SemProM-Architecture*

generation RFID, and build on top of them. He outlined a "layered architecture" where more investment brings additional services. At a basic level, you might have technologies with simple product memory but no information processing (such as barcodes and simple RFID); at a higher level, you might have product tags with sensors, data storage, and the ability to communicate with other tags; at the most advanced level you might have "smart" items with semantic integration of data and semantic query processing.

In short, Prof. Wahlster said, future manufacturing is likely to be the crucible of the IOT. This match will produce killer apps, like semantic product memory, and drive new innovations in how we do business across the entire product lifecycle.

## The Discussion

Once again, the group enjoyed a spirited debate and discussion. All agreed that future manufacturing is a fertile field for the IOT.

Yet many questions clouded this vision. Are information systems sufficiently integrated to realize the IOT? Are sensors reliable? What about real-time capacity? How to connect the IOT with processes such as pull or push systems? What changes lie ahead?

Ms. Murphy-Hoye of Intel asserted that the IOT alters our fundamental approach to manufacturing. "As soon as you start to use RFID or sensor networks, anything like that, you really change the nature of what it means to represent the factory, to interact in the factory, and to actually manufacture something," she said.

Intel has already begun to explore the factory of the future. For example, Ms. Murphy-Hoye and her colleagues at Intel have studied the "ethnography" in their factories. They *become* the product and learn what happens to the product itself. This form of inquiry is very different from the traditional perspective, which usually focuses on equipment (especially in a capital intensive industry). Technology that allows memory of the product lifestyle enables this sort of approach.

"When you turn the equation upside down and you start to look at things from a different perspective you really begin to see the semantic gaps between how to logically represent what happens there and what physically happens," she said. "You start to think about this concept of self-organization in the factory, where the product finds its own way, determines its own destiny. This allows for mass customization without creating factories that are specialized and customized."

Such a new perspective obliges us to rethink the nature of manufacturing, said Ms. Murphy-Hoye. Products will evolve just as software does—and the IOT will show us how this happens. Similarly, it will show us what happens when these products move through the value chain and interact with other elements in the IOT and people.

"How is that evolution of the product really a part of manufacturing?" asked Ms. Murphy-Hoye. "How do you rethink the definition of manufacturing when you have this type of product and environment?"

The notion of a product lifecycle also opens up new areas for innovation. Dr. Guilherme Vaccaro of UNISINOS Brazil said the IOT also provides more information about how a product is used—and this could help design.

"If you could just grab the information from a user's product, you could perhaps improve the ability to do better designs and be more cost effective," said Dr. Vaccaro.

The IOT also opens the door to opportunities in services, said Prof. Dr. Martii Mäntylä from the Helsinki Institute for Information Technology. If self-monitoring and self-diagnosing systems call attention to problems, it creates a demand to fix those problems. "So perhaps the service content of products could be much larger in the future because of that, and, if so, that is a major economic possibility," he said.

The IOT also forces us to rethink the role of humans. Dr. Matthias Kirchmer from Accenture emphasized the role of people and collaboration in the IOT. Echoing the thoughts of Mr. Bullotta, he suggested we should structure manufacturing processes based on the roles and skills of the personnel.

"You create something like communities of products or communities of sub-products, where you can basically align the different groups of people involved based on the final goal, based on the solution and the product they want to deliver to the market together," said Dr. Kirchmer.

Similarly, this idea could be expanded beyond the individual company. These communities might include suppliers and customers as well. "You have communities around your entire value chain," he said, "and, with that, you get people connected and enable them to have the flexibility to manage the chaos."

Unfortunately, Dr. Kirchmer added, at this point, the systems are not ready for this level of integration. What are the requirements for this integration of systems? Dr. Kirchmer said there is no general answer and it must vary by industry. "The role of that real-world integration and the role of the Internet of Things, again, should not be defined in general for the manufacturing of the future because there may be a manufacturing of the future that looks very different," he said.

Prof. Heuser of SAP emphasized that these innovations must be integrated with existing enterprise applications. Otherwise we will have the kind of nightmare we saw in the past.

"If you end up doing a one-off again, then you are back 30 years where you had all your individual software packs developed and then you ended up in a mass of maintenance," Prof. Heuser said.

How should we begin to integrate? "There are many levels of possible integration between things and processes, and also, there are very big differences between the different manufacturing industries," said Dr. Vaccaro. "It's difficult to generalize."

Dr. Vaccaro added another note of caution. We also must be careful about generalizing, because the costs of labor and technology vary a great deal in different parts of the world. In countries with cheap labor and less technological advances, there would be less incentive for smart factories. "The integration of the Internet of Things should take into account the reality that in some countries a machine costs much more than 10 people," he said.

The question of human labor raises other issues. As Dr. Vaccaro noted, "In the future, production processes will be highly automated, but you have to find a way to improve the value of the human labor and this could be done by the Internet of Services, for instance."

What other forces will drive the IOT? One is as old as business itself. "It's an opportunity for companies to excel in comparison to their immediate competition," said Prof. Günther from Humboldt University. "It's an entrepreneurial decision. It's a risk, to dive into it, but some will be able to excel this way."

Ms. Murphy-Hoye agreed. "Competitive differentiation is what is going to drive this," she said.

Once again, the cautionary tale of Wal-Mart's RFID system came up. Forum moderator Dan Woods asked, if Wal-Mart didn't do it right, can anybody?

"Wal-Mart has set RFID applications back quite a large degree," said Mr. Woods. "The brand impression of Wal-Mart is that they're very, very good at logistics and IT, and if they can't make this work for them, for whatever reason, even if it's their fault, it's certainly going to put a black mark on the Internet of Things."

Luckily, others did not see the Wal-Mart precedent as an insurmountable barrier.

According to Ms. Murphy-Hoye, the problem was not the technology; the problem was Wal-Mart's use of the technology. Others have had more success by approaching this technology from the perspective of product lifecycle and competitive differentiation.

"If you look across the world, what's happened with RFID has not been about slap and ship, it's been about the value of inherently transforming our products and our own manufacturing processes," she said. "We're getting great benefit from this investment approach, one that Wal-Mart didn't provide for its suppliers."

Speaking of Wal-Mart...Prof. Wahlster issued a plea: could we please stop talking about that example?

"I think the Wal-Mart example is really horrible because Wal-Mart was not about the Internet of Things," he said. "This was just tagging stuff... We are now beyond RFID. If you put an RFID chip on something this is not the Internet of Things because there is no Internet connection. We should drop the Wal-Mart counterexample because Wal-Mart was not about the Internet of Things, but simple tagging of products with first generation RFID."

Is the IOT worth getting excited about?

At one point, Mr. Woods observed that the IOT didn't generate as much excitement among IRF participants as, say, the discussion of mobile did the previous year. "I'm getting the impression for both the Future Factory and the Internet of Things in general that they're a good direction," said Mr. Woods, "but I'm not getting that sense of urgency."

"I don't know that it's about urgency," responded Ms. Murphy-Hoye. "Inevitability is what we're talking about. This is happening and it's changing."

Again, Prof. Wahlster stepped in. He saw a pressing demand.

"I really disagree that there's no urgency in manufacturing," he said. "Worldwide in the manufacturing industry the hottest topic right now is the digital factory based on IP technologies."

## Setting the Agenda

By the end of the 2009 IRF, we had many pages of observations and questions, all of which provided an agenda for subsequent research. In past years, the authors have woven this follow-up research into the conference proceedings. This year, we take a fresh approach and devote the second half of this book exclusively to answering these questions. We divide the rest of this book into three basic areas: building the IOT, using the platform, and forecasting for growth. We identified key questions—some of which will be familiar, others new—that will guide these discussions.

# IRF 2009

## Part Two

The second part of this book explores unresolved questions that arose during the conference. Inevitably, the IRF produces more questions than can be answered in the short amount of time available. This part of the book reflects research done after the conference. It explores how the Internet of Things will be built, used, and grow.

# Standards

# 6

In the IRF 2009, almost every discussion inevitably turned to the question of standards. This reflects the importance—and difficulties—of standards for the development of the Internet of Things. Standards are essential for interoperability and achieving the network effects that potentially make the IOT so powerful. To realize the vision of the IOT as a next-generation Web where billions of objects are networked, these standards must be interoperable, scalable, and support future innovation. Ideally they should work with existing systems and future ones.

Yet the IOT is dogged by many unresolved questions: what will the standards be? Who should set them? Will there be one standard or many? Will standards be dictated by some central organization or arise from the tumultuous marketplace? Can the players come together and create standards that allow all the potential parts of the IOT to work together in harmony? How can we set open standards yet still ensure that vendors can protect intellectual property and have the freedom to innovate?

## Why Standards?

While almost all observers agree that technical standards are essential to realizing the IOT, consensus on how and what standards to set has been slower to come. The emergence of clear standards has been slow due to the role of many industry players and the existence of competing standards organizations.

As a practical matter, when speaking of the IOT, standards are sets of rules, guidelines, and principles governing how things should be described and implemented in software. Ideally, broad standards ensure interoperability between vendors and are distinct from proprietary standards or intellectual property that belongs to one company.

As an example, let us look at RFID standards. RFID stands out as the preeminent technology of the IOT (other technologies are on the horizon, but not yet deployed). With RFID, each object has its own unique identifier. In contrast, barcodes usually are used to identify classes of products, not individual items. This unique RFID identifier allows objects to be read from a distance with wireless technology, thus providing real-time identification and tracking of individual items. RFID tags are now produced by vendors and deployed across the world, yet there is no global consensus on standards. As a result, we are witnessing a proliferation of identity codes for both active and passive RFID. RFID has been implemented in various ways by different manufacturers in different regions of the world.

According to Dr. Zoltán Nochta, Deputy Director of the SAP Research Center in Karlsruhe, Germany, there are about 20 variants of RFID technology with different bandwidths and frequencies. Even barcodes have multiple standards. This heterogeneity continues throughout the stack for every kind of functionality.

Dr. Nochta threw up his hands in frustration.

"Multiple industry organizations and national as well as international standardization bodies, including EPCglobal, address different aspects of identity code management. These organizations define

**Dr. Zoltán Nochta** is Deputy Director of the SAP Research Campus-based Engineering Center (CEC) located in Karlsruhe, Germany. At CEC, he manages teams that work on multiple research topics including smart items, IT security, and privacy protection, as well as smart grids and future energy markets.

Dr. Nochta is also the SAP Research Intrapreneur for the Internet of Things and drives an initiative to help position the company as a service provider in this emerging business. He was also responsible for preparing the content and discussion topics for the IRF 2009 in Dresden.

Dr. Nochta received his Ph.D. at the University of Karlsruhe where he conducted research in the area of secure access control and cryptography in distributed systems.

their own numbering schemas, data formats, and the related technology stack for their respective business goals and processes," he said. "Everybody's doing his own stacks and doesn't really accept the meanings and the mechanisms of the others. Picking random numbers, for example, as object identifiers seems to me in many cases just to be the better approach, because it would make at least the process of unique identification easier and simpler.

"Multiple standards create confusion. Take the example of exchanging data between an automotive manufacturer and a retailer of car parts. The automotive and retail industries use different standards to describe the same part. The software has to support both forms of numbering, map and remap, transform and retransform to get the complete tracking and tracing history of a certain part. And all of that makes your software very complicated," he said. "It's a mess."

Standards are particularly important for data interface protocols (as we move higher up the stack into software, there is more incentive for people to add proprietary elements). Unfortunately, we are seeing waves of instrumentation coming without interoperability and little thought about distributed architecture. According to Ms. Murphy-Hoye from

Intel, we're not yet creating an Internet of objects as much as a "mess of instrumentation."

There is a general consensus on one thing—we need more universal standards. Standards bodies are under pressure to define global requirements for identity and naming schemes to ensure interoperability. These standards bodies must define matters such as the networking layers of the infrastructure, the functionality of device interfaces, data formats and information codes, naming, addressing and identification, middleware, and interoperability.

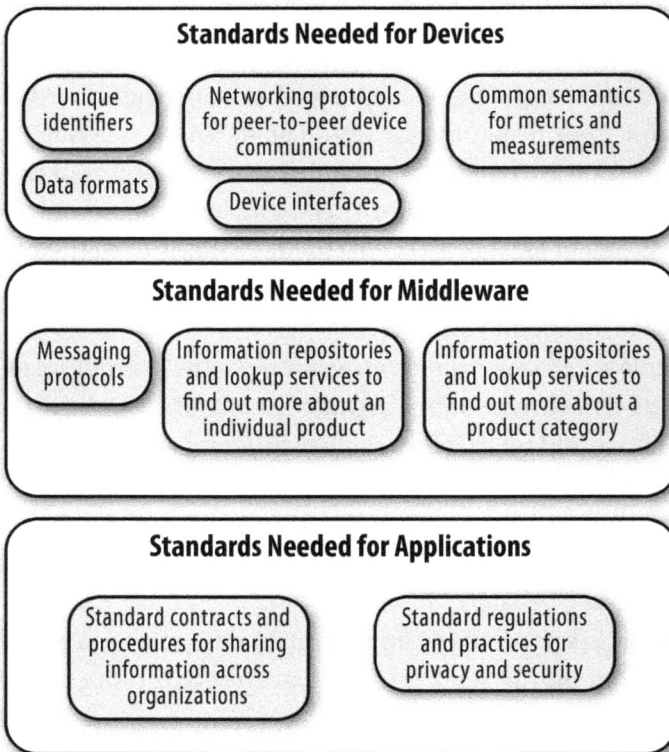

**Standards Needed for Devices**

- Unique identifiers
- Networking protocols for peer-to-peer device communication
- Common semantics for metrics and measurements
- Data formats
- Device interfaces

**Standards Needed for Middleware**

- Messaging protocols
- Information repositories and lookup services to find out more about an individual product
- Information repositories and lookup services to find out more about a product category

**Standards Needed for Applications**

- Standard contracts and procedures for sharing information across organizations
- Standard regulations and practices for privacy and security

*Figure 6-1. IOT Areas Needing Standards*

Yet standardization efforts remain uncoordinated. As one European Commission report noted, "While much is being done, worldwide efforts are still quite fragmented and decisions are sometimes taken by ad-hoc organizations which do not necessarily follow the principles guiding EU standards organizations."[1]

Standards efforts often bog down because each industry tends to become attached to its own parochial approach. Industries cling to their own standards and do not want to adopt those of another group. As a result, efforts to broker universal standards become drawn out and frustrating affairs. As Mr. Bullotta of Burning Sky Software, says, "Standards are like toothbrushes. You know you need one, but you don't want to use someone else's."

How to clean up the mess? This problem could be solved if everybody could agree on some kind of common denominator. "It would be doable," Dr. Nochta says with a laugh, "but nobody is doing it for some reason."

## The Problem

According to Dr. Jochen Rode, Development Architect and Smart Items Research Program Manager at SAP Research Center Dresden, existing RFID applications have solved only a small portion of the challenges that face us. Today, a number of standards exist that define the frequencies to be used by RFID readers. However, beyond this low-level integration not much else is generally agreed upon. For many of the larger IOT issues, namely globally unique object IDs, ID resolution, and semantic data formats, no universally accepted standards exist yet. Typically, we see an array of competing and often proprietary formats.

Today, RFID largely remains confined to closed-loop scenarios, such as container tracking and tracking items in a factory. The more interesting and promising applications will come from more open scenarios where

---

[1] Commission Of The European Communities, "Future networks and the internet: Early Challenges regarding the 'Internet of Things'," *http:// ec.europa.eu/information_society/eeurope/i2010/docs/future_internet/swp_ internet_things.pdf*

**Dr. Jochen Rode** has a German diploma in business and information technology, and earned a master's and Ph.D. in computer science. His background ranges from networking and software engineering to web application development, human computer interaction, and manufacturing. Dr. Rode joined SAP in 2005 as a senior researcher. In his current position as Development Architect and Smart Items Research Program Manager, his primary interest is applying smart items technologies (such as RFID, sensors, and embedded systems) for SAP's customers. Dr. Rode also leads SAP Research's R&D for the manufacturing domain and manages the SAP Research Future Factory Living Lab *(www.sap.com/futurefactory)* at Dresden. In this context he leads a number of projects related to device integration, machine data acquisition, and control.

partners share information with each other and with consumers. "For that," says Dr. Rode, "you need standards."

A good example for that is the EPC (Electronic Product Code) standards. It includes elements such as GTIN (Global Trade Item Number), SGTIN (Serialized Global Trade Item Number), classical Internet DNS and Object Naming Service infrastructure and EPCIS (Electronic Product Code Identification System of EPCglobal). This allows storing and accessing data such as movements of goods or status changes. Since the release of EPCglobal standards, open scenarios have become more commonplace.

But Dr. Rode notes that even EPCglobal falls short of what is needed for a general IOT infrastructure. Many industries do not want to use EPCglobal for several reasons—most of them nontechnical. Some express their perception that EPCglobal does not fit their needs; others already have alternative solutions that are costly to replace, fear loss of control, are unwilling to pay for getting IDs assigned, or simply suffer from the "not-invented-here" syndrome. While EPCglobal addresses the basic needs for tracking goods movements, Dr. Rode says it falls short in defining deeper semantics such as storing the temperature curve in a cool chain. EPCIS enables extensibility but still requires collaborating partners to agree on a semantic standard. "For those reasons, it is by no means sure that EPCglobal will become the future IOT backbone," he says.

Occasionally, the Universal Unique ID (UUID) standard is mentioned as an alternative to a centrally and privately managed ID system like EPC. While it has the advantage that IDs can be generated in a decentralized manner, we confront a different set of problems. "You can have billions and billions of unique IDs, but who provides the lookup?" asks Dr. Rode. "If you don't store all information locally on the product, you need to ask somebody, 'Could you provide me the data for that particular object?' And who do you ask? Of course everybody wants to be the source for that data. Some of the major players want that role, but there is really no solution for this problem right now."

Technically, Dr. Rode says EPCglobal addresses most of the needs to realize the vision of the digital product memory and global goods tracking. If the nontechnical issues can be addressed, EPCglobal, or a similar infrastructure, may become part of the future IOT backbone. However, the IOT is more than just about exchanging static product information. Most interpretations of the IOT include the ability to get dynamic sensor data in real time from anywhere in the world—another aspect currently not addressed by EPCglobal.

## Standardization Initiatives

While there are many standardization initiatives underway, such as EPCglobal, ISO, IPSO Alliance, and others, each effort is unique in terms of the scope and goals for the standards.

## EPCglobal

EPCglobal has its roots in the Auto-ID Center, a project founded in 1999 at the Massachusetts Institute of Technology (MIT), which worked with industry partners to develop RFID standards. The Auto-ID Center produced the Electronic Product Code (EPC), a numbering scheme for physical goods for supply chain applications (followed by standards for communications between transponders, scanner hardware, and information systems). In 2003, the technology was handed off to EPCglobal, a non-profit organization responsible for developing standards. The EPCglobal standards system enjoyed a major boost when it was adopted by large retail chains such as Wal-Mart, Tesco, and Metro and the US Department of Defense.

Today EPCglobal is a subsidiary of GS1, a global not-for-profit standards organization. The GS1 System is an integrated system of global standards that provides for accurate identification and communication of information regarding products, assets, services, and locations. EPCglobal is a global standards system that combines RFID (radio frequency identification) technology, existing communications network infrastructure, and the Electronic Product Code (a number for uniquely identifying an item). This system enables immediate and automatic identification and tracking of items through the whole supply chain. GS1 also has standardization initiatives in other areas related to supply chains including barcodes, electronic business messaging, and data synchronization.

EPC started with retail and consumer products and branched into other sectors such as healthcare, consumer electronics, transportation, logistics, aerospace, and defense.

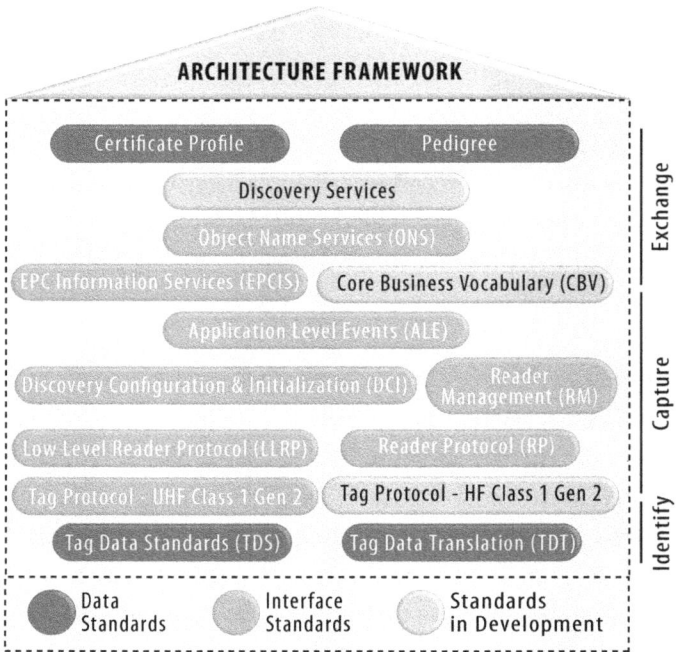

**ARCHITECTURE FRAMEWORK**

Certificate Profile          Pedigree

Discovery Services

Object Name Services (ONS)

EPC Information Services (EPCIS)    Core Business Vocabulary (CBV)

Application Level Events (ALE)

Discovery Configuration & Initialization (DCI)    Reader Management (RM)

Low-Level Reader Protocol (LLRP)    Reader Protocol (RP)

Tag Protocol - UHF Class 1 Gen 2    Tag Protocol - HF Class 1 Gen 2

Tag Data Standards (TDS)    Tag Data Translation (TDT)

Exchange    Capture    Identify

Data Standards    Interface Standards    Standards in Development

*Figure 6-2. EPCglobal Standards*

## EPCglobal Network Components

The EPCglobal Network includes seven standards-based hardware and software components. Although the EPCglobal approach is not universally adopted, it provides an example of key areas to be addressed by any standardization effort.

- Electronic Product Code (EPC)—Unique number to identify specific items such as containers, pallets, cases, or single items
- EPC Tag—Radio frequency tag containing a microchip with the item's EPC along with an antenna to convey information to a reader
- EPC Reader—Device that detects tags and relays their EPC numbers to middleware
- EPC Middleware—Software for managing data from EPC readers
- Object Naming Service (ONS)—Network resolution services that direct EPC queries to a location where EPC information can be accessed by authorized users
- EPC Information Services (EPC-IS)—Information services for the storage, communication, and dissemination of EPC data in a secure environment
- Discovery Service—Mechanism for securely locating all events and information for a given EPC

### The EPCglobal Standards Stack

The EPCglobal architectural framework involves multiple standards related to RFID and related technologies. This framework identifies 15 standards for data, air interfaces, and other areas under development.

Gay Whitney, standards director of EPCglobal, describes the standards as "componentized." This means that each standard is designed to work either by itself or in conjunction with other standards. The standards focus on air interfaces and data exchange, and can work with other numbering systems besides the GS1 numbering system.

### Setting Standards

According to Ms. Whitney, EPCglobal follows a user-driven process for development of standards and has working groups devoted to standards development. EPCglobal is a member-driven organization of both users and solution providers. Participation in the standards-setting process is mostly limited to paid subscribers. "We spend a lot of time on user requirements," says Ms. Whitney. "We bring the industries together in this joint requirements development, which then leads to standards development."

Under this approach, EPCglobal first seeks to understand the needs of the industry. Once they obtain a clearer picture of the market, GS1 determines whether that sector requires another set of unique standards.

According to Ms. Whitney, EPCglobal tries to make standards royalty-free whenever possible. Anybody is free to use these standards, regardless of whether or not they are a subscriber. (That said, private companies do retain the right to protect intellectual property that they develop on top of these standards.)

"I think the perception for many years was that there was an effort to control the IP, and therefore control the marketplace," Ms. Whitney says. "We work toward ensuring that the core functionality will be interoperable. We simply try to create a foundation of base, core functionality that could be easily adopted and create an environment where the technology could be most widely deployed."

EPCglobal has faced some criticism and some misperceptions. For example, some believe that it is a private company or that its standards are proprietary. "The standards can be used by anyone," says Marisa Jimenez, Public Policy Director Europe of GS1. "The standards are about interfaces. They're not software, they're not hardware. There was a little bit of confusion out there, an unintended perception, that we were selling a technology or a final product, which is not the case."

EPCglobal collects subscription fees based on the size of the company and home country. This subscription ensures that the organization can effectively support members' needs in key areas like public policy, research through the Auto-ID labs, and standards development.

"We, as a standards-setting organization, may not have all the answers; our greatest strength is in bringing people together to discuss what is needed as technologies develop and improve over time," says Ms. Whitney. "Through our community-driven discussions we work toward finding the most appropriate solutions that will serve global needs most effectively."

This dynamism poses some challenges for setting standards. "At some point, you have to put a stake in the ground and standardize," she adds. "We feel the componentized approach is the right approach. As you identify the most critical components that need to be standardized in order for this vision to be realized, you can address each based on the priorities of the community."

Typically, a group of members will clamor for a standard (or standards) to support a specific scenario or use case. If there is a demonstrable need for additional standards, EPCglobal will step in and shepherd the process of setting new standards. Ms. Whitney expects this process to accelerate as the economy improves and enterprises invest more in these new technologies.

The EPCglobal approach is different from that of ISO. There is some duplication of effort, yet the two groups have coordinated their approaches to standards. In 2006, the two organizations formed the EPCglobal-ISO Committee to coordinate efforts on the UHF air interface. The EPCglobal community submitted the Gen2 standard for the UHF air interface to ISO for ratification and it was adopted as the ISO 18000-6C standard.

"We developed an approach that we have used to ensure that as technical standards progressed in EPCglobal, this work was shared with the comparable work efforts in ISO," says Ms. Whitney. "And we extended that information sharing beyond the air interface and collaborated on the software standards as well."

### ISO

Meanwhile, the International Organization for Standardization (ISO) took its own tack toward RFID. ISO has taken a more generic approach and its standards cover the areas of technology such as air interface standards, data content, conformance and performance, and application standards. The ISO RFID standards are generally more application independent. That said, there is some overlap. As mentioned, EPCglobal's Gen2 standard for UHF was adopted as an ISO standard (18000-6C).

### IPSO Alliance

The IP for Smart Objects (IPSO) Alliance seeks to promote the Internet Protocol (IP) as the networking technology best suited for connecting sensor- and actuator-equipped or "smart" objects.

The IPSO Alliance is an open group of member companies such as Cisco, Intel, SAP, and Sun Microsystems. The group formed in response to a market that was becoming increasingly fragmented by many proprietary ad hoc solutions for connecting these devices. The alliance seeks to bring together key players, such as vendors of chips, software, and end systems, to establish IP as the way forward to connect smart objects and sensor networks.

The IPSO Alliance argues that IP has proven to be the most efficient and scalable networking technology. IP is a proven, ubiquitous standard that should serve as the bedrock of the IOT. The alternative is a patchwork of proprietary protocols with no guarantee of scalability or interoperability and complex gateways. IP can fulfill the requirements for IOT devices for low power, restricted memory, rugged surroundings, and remote unattended devices.

The IPSO Alliance does not seek to define new protocols. Instead, it works with international standards organizations such as the IETF, ISA, IEC, IEEE, and W3C and utilizes the standards developed by them.

The group argues that using IP will simplify the development of new applications because it will utilize a familiar paradigm for programming and networking, existing protocols, and existing tools.

Another issue is addressing. Each computer has a unique IP address, either dynamically at the moment of connection or for a period of a day or so, or (for all intents and purposes) a fixed or static address, like that assigned to a web server or DNS server. The current version of IP, version 4, allows for 4.3 billion unique addresses. A few years ago, that was deemed adequate, but now there are only 1 billion addresses left with more people and devices coming online by the minute. The successor, IPv6 has 340 billion address slots that seem (at least for now) to offer limitless web access and unbreakable encryption levels.

IPSO supports both IPv4 and IPv6. The Internet of today largely rests on IPv4 and thus it is critical for the IOT to provide IPv6 interoperability. Yet IPSO also recognizes that IPv6 offers features such as an immense address space, address auto-configuration, and header compressibility that will be desirable in the future. It recommends that both standards be recognized to ensure interoperability with existing systems and the ability to adopt better ones.

## Standards in Inaction: The Auto Industry

**Prof. Dr. Elgar Fleisch** is Professor of Information and Technology Management, ETH Zurich and University of St. Gallen (HSG). His research focuses on the economic impacts and infrastructures of ubiquitous computing. In the Auto-ID Lab he and his team have developed, in concert with a global network of universities, an infrastructure for the Internet of Things. In the Bits-to-Energy Lab, which he co-chairs with Prof. Mattern of ETH Zurich, he investigates and designs technologies and applications to save electrical power and water. In the Insurance-Lab. Prof. Fleisch, together with Dr. Ackermann of HSG, drives forward technology-based innovation in the insurance industry. All research projects are joint efforts of industry and academia and the findings have been published in over 200 scientific journals and books.

Prof. Fleisch is a co-founder of several startups (for example, Intellion, Synesix, Coguan, Dacuda, and Amphiro) and serves as a member on multiple management boards, and academic steering committees.

The adoption of RFID has taken considerably longer than predicted and much of the delay can be attributed to the lack of standards. An illustrative example can be found in the automotive industry. This case is described by Prof. Dr. Elgar Fleisch, Professor of Information and Technology Management, ETH Zurich and University of St. Gallen (HSG), and his colleagues Patrick Schmitt and Florian Michahelles, in the paper "Why RFID Adoption and Diffusion takes Time: The Role of Standards in the Automotive Industry." Their observations ring true across many sectors well beyond the auto industry and highlight the general problems facing the IOT.

The authors cite three reasons for the slow rollout of RFID:

- Skepticism about EPCglobal and the capability of the Electronic Product Code for the needs of that particular industry

- Protracted and unresolved standardization processes

- The lack of a standardization mandate by a dominant supply chain partner or body

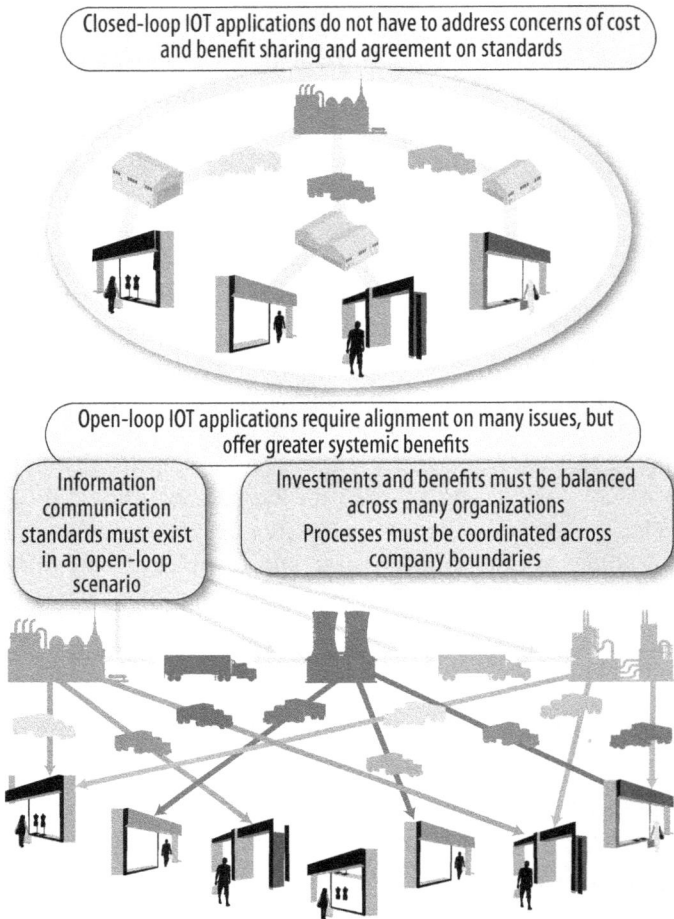

*Figure 6-3. Open-Loop versus Closed-Loop IOT Applications*

In theory, RFID should be a very attractive technology for the auto industry. Over the last 15 years, the automotive industry has moved away from the classic tools of data transmission like phones, faxes, and paper and shifted toward Electronic Data Interchange (EDI), Enterprise Resource Planning (ERP), and barcodes—all of which have increased the volume, quality, and flow of data. RFID seems a natural evolution of this trend. After all, RFID has the potential to enhance data capturing and processing, increase visibility across the supply chain, and reduce cycle times.

Most RFID applications within the auto industry are confined to closed-loop scenarios involving one company or perhaps a small number of trusted partners at specific sites. In these cases, the RFID transponder is often reused or remains within a limited area, such as a factory. Other closed-loop scenarios might include RFID tool management, assembly control, or a Kanban system within a plant. Collaborative closed loops could include container management or the management of transportation racks between an OEM and a first-tier supplier. Manufacturers such as DaimlerChrylser, BMW, and Volkswagen are already using or testing RFID systems in closed loops.

A natural progression of these RFID technologies would be open-loop applications where the RFID transponder remains on the object as it makes its way through the supply chain. But open-loop applications remain rare. To be sure, the rollout of new technologies typically begins with pilots in closed environments where RFID tags can be reused and processes repeated. But it also reflects the lack of an agreement on RFID standards, especially data standards and unique identifiers like the Electronic Product Code (EPC) that can be used across enterprises.

Widespread lack of clarity and consensus has slowed adoption. The auto industry has been reluctant to embrace EPCglobal, which is dominated by the retail sector. The automotive industry largely conforms to ISO standards and probably expects clearer ISO standards on RFID. As Prof. Fleisch says, "The European automotive industry is still a bit skeptical regarding technology-driven organizations like EPCglobal. They like to base their systems on standards offered by classical standards bodies, such as ISO."

Powerful trade groups and associations in the auto industry have claimed the right to standards development and see it as one of their core competencies. The US-based Automotive Industry Action Group, founded by DaimlerChrysler, Ford Motor Company, and General Motors, has promulgated RFID standards for tire and wheel labels. The Association of the German Automotive Industry has issued recommendations for using RFID in container management processes. Odette International, which represents automotive companies and associations in six European countries, launched its own RFID standards initiative.

The auto industry was hardly alone in pursuing its own standards. The pharmaceutical industry wanted RFID tags to include lot/batch numbers, expiration date, and the National Drug Code. The electronics industry wanted harmonized tariff codes and serial numbers. More wish lists came from other industries: logistics, airport transport, aeronautics, telecommunications, and many others.

Multiple standards might be tolerable if goods stayed confined to certain industries. In reality, they do not. A tire may be coded one way in the auto industry and another way in the retail sector. To fully realize the potential value of the IOT, each player needs an application program, middleware, and reader protocols to read memory data. If one partner adds information, the whole network needs an update.

In short, industries have taken a bottom-up approach to standardization based on local needs. As Prof. Fleisch and his colleagues note, this approach has curtailed RFID adoption in the auto industry and many other sectors. As they conclude, "A top-down strategy for RFID standards development would have been more promising: First, to adopt a generic standard with a 'one-fits-all' concept and second, to look into detailed processes and applications and then to adjust these processes to the overall concept."

## Industrial Islands of Standardization

The auto industry illustrates trends that are occurring in many sectors. Similar dynamics are playing out in pharmaceutical, chemicals, electronics, and other industries.

Prof. Dr. Oliver Günther of Humboldt University, Berlin, Germany sums up the standardization debate: "It's essentially about power and money."

"They're not so different from the standards strategies you see in other domains," says Prof. Günther. "Standards are really about power, so when you establish standards you have different stakeholders and they try to maintain their power base, and that's what makes standardization processes very difficult."

The EPCglobal standard was developed for retail. On the one hand, it offered a proven standard that other industries could join so they didn't have to reinvent the wheel. On the other, it was designed to meet the

**Prof. Dr. Oliver Günther** is Dean of the School of Business and Economics at Humboldt University in Berlin, Germany. He also directs Humboldt's Institute of Information Systems and its Interdisciplinary Center on Ubiquitous Information. Prof. Günther has held visiting appointments at the European School of Management and Technology, Tsinghua University in Beijing, ENST Paris, UC-Berkeley, the University of Cape Town and, most recently, at SAP Labs in Palo Alto. He is on the Advisory Board of SAP Research and served as an IT strategy consultant and board member to numerous government agencies and high-tech companies. His current research interests include business applications of social networks, RFID architectures, IT productivity, and security and privacy in ubiquitous computing.

needs of retail and might be less suitable in other industries and required others to pay EPCglobal Consortium to administer it.

The car industry has been hesitant to adopt the EPCglobal standard. What's in it for them? The benefits remain unclear. "Is it really important for them that their carburetors are numbered in a way that is compatible with what EPCglobal does for the shirts and the coats that their members produce?" asks Prof. Günther. "It's not so clear whether the cost benefit situation really points toward the election of global standardization."

As a result, Prof. Günther expects industry-specific standards to prevail for the near future. Some industries will be too strong and independent to come under the umbrella of a global association. "I think for quite some time we will have industry-specific standards that grow out of legacy numbering schemes," he says. "There will be transfers between these different industry specific solutions."

Some have suggested that industries will develop their own islands of standards. Prof. Günther extends this metaphor: he believes industries ultimately will group themselves into "continents of standardization." Clusters of industries will share standards. In the inter-company application space, companies like SAP can play a valuable role by helping to broker standards and create platforms that allow interoperability and cooperation between players.

Once profitability becomes clear, solutions and applications will emerge. According to Prof. Günther, players have failed to "think big." They have remained bullish on certain technologies without realizing how much the overall IOT might transform their business processes.

"I don't think the standardization will be a total roadblock," he says. "You just have to show people how they can get their money's worth."

Many prominent experts agree. Prof. Min of Fudan University predicts that these continents of standards will merge over time.

In the near future, Prof. Min says industries will devise their own standards according to their requirements. Standards will compete, some will rise to the top, people will combine the best elements, and eventually we will arrive at global standards. Until then, standards will be connected by translators or routers. Some manufacturers may put more than one standard on their products.

Prof. Min points out that this is just what happened with the Internet. At the beginning, there were many different networking interfaces. Eventually, the world adopted Ethernet and TCP/IP.

**Dr. Ralf Ackermann** holds a Ph.D. in computer science from the TU Darmstadt where he worked in the field of IP telephony services in heterogeneous environments at the Multimedia Communications Lab (KOM). After working as team leader for the ubiquitous computing research group at KOM and founding a startup company, he joined SAP Research in July 2006. There he worked as a project lead for various research initiatives in the fields of RFID and semantic technologies. In addition Dr. Ackermann has a strong interest in systems engineering and heterogeneous systems, ranging from microcontrollers and sensor nodes over OS kernel and device driver programming to building large industry grade (communication/real-time) systems. He has a number of referenced publications and works as a reviewer for the EU.

"I think at the very beginning there will be various different standards," says Prof. Min. "Industry sectors may have their own standards, and every country may try to have its own standards. But eventually I think there will be a global standard for the description of the objects."

Dr. Ralf Ackermann, Project Lead, RFID and Semantic Technologies, SAP Research, also foresees a convergence of standards. He understands the desire for end-to-end standards, yet he sees practical value in heeding the example of the Internet and taking a more modest approach of aiming for interchangeable building blocks. Inevitably, there will be need for niche standardization in certain arenas. He recommends that we keep building blocks like IP version 6 and build the IOT on top of them. In the long run, Dr. Ackermann predicts just one communication infrastructure and code for both the classical Internet and the IOT.

"I don't think that in the long-term future there's going to be a separation between the consumer Internet, where it's about streaming information, and the IOT," he says. "Those two things actually belong together."

## Taxonomies and Ontologies

The question of standards also raises related questions about definitions. How do we define universal meanings for terms that are essential to our businesses? Does the weight of a product refer to gross weight or net weight? The weight of a pallet or a single item? We need to build ontologies that define the things that are essential in our businesses.

"We really need to build ontologies that are probably going to be domain-specific," says Dr. Rode. "We need agreement on what these things mean. And ontology doesn't have to be a complicated concept—it could be a Word document that says, 'If I say I have the weight of a product, it's always the net weight, minus all the packaging.' Unfortunately, there's not going to be an overarching ontology for everything that's in the world, because the domains are way too different."

In each field, leaders will need to get together and establish taxonomies of concepts, give those things names, and standardize them. "That has to happen in each industry—the semiconductor industry, the food industry, and so on and so on," says Dr. Rode. "You have to do this over and over. I think there is no easier way."

## Legacy Systems

Standards not only must give us a path forward; they must also connect with the existing systems. Mr. Bullotta emphasized that any new standards need to be compatible with legacy systems. He estimated that web developers spend one-third to one-half of their time grappling with the incompatibilities of different browsers and standards. "If you look at the standards that have really been impactful, they actually fall in the lowest common denominator group," he says.

## Making Peace with Heterogeneity

Unlike the Internet, there is no single global set of standards for the IOT. According to Prof. Fleisch, there never will be.

"I strongly doubt that we ever will have one single standard for every single object that we want to access through the Internet via the additional layer of the Internet of Things," he says. "Maybe it is a good vision to say we need one global standard that can address each and every thing via the same standards, but I don't think that's true."

Is the broad standardization achieved by the Internet an exception or a model? The IOT more accurately reflects the physical world—one that is vastly more complex than the old Internet, and one that is not highly standardized. Like the natural world, the IOT must reflect this diversity and its applications must adapt to local environments.

Standards inevitably will emerge—the IOT would not be viable without them—but they will be balkanized. Standards will arise out of certain industries, like retail, automotive, or telecommunications. Once a standard captures a large share of the market, the cost drops dramatically and more industries are likely to adopt that standard. This is precisely what happened in the retail sector: EPCglobal became the de facto standard, largely due to the mandates of mega retailers like Wal-Mart and Metro. This standard has spread to related industries, such as textiles.

Dr. Nochta proposed an architecture of layered standards. People could choose to affiliate with this or that stack, depending on their particular needs. "You can have a heterogeneous technology landscape and many standards within the same layer," he says. "But what is not allowed is having different interfaces between the layers; there must be one."

Each layer would be tailored to a specific industry; interfaces would allow these layers to communicate. "You can have your vendor-specific products and technologies," says Dr. Nochta, "but at the end they work together."

Such an approach worked for the Internet. There are many different standards and protocols but ultimately they all fit together. The Internet was a truly general-purpose stack that was developed for a purely functional purpose. Now almost every standard-setting effort is heavily influenced by the needs of particular industries. It seems unlikely that the IOT will have a general-purpose set of standards. Instead, says Dr. Nochta, we must learn to live with many standards for the foreseeable future.

"Because of this siloed thinking, don't expect that there could be something like a unified Internet of Things—not by a long shot," he says. "There will be fragments of the Internet of Things, which are not really connected to each other."

That leaves vendors who serve all these industries in a quandary. According to Dr. Nochta, companies like SAP, IBM, or HP inevitably ask themselves whether they should invest in small spaces with a few customers who share a certain standard. In the end, these vendors may decide it's not worth adopting the particular standard of this niche or that niche. Instead, they may decide to develop all-purpose platforms and rely on connectors and consultants to adapt solutions on a case-by-case basis. "Everybody is doing the same thing in the face of fragmentation," says Dr. Nochta. "They adapt themselves to this fragmented, non-unified situation."

Burkhard Neidecker-Lutz, Technical Director, SAP Research, argues that we should make peace with the lack of standardization for another reason—heterogeneity makes sense from a technological perspective. He believes that universal standardization is a nirvana that we will never attain.

"There will not be a convergence on just one set of standards as we've seen with the rest," he says. "So basically, deal with it. There will always

**Burkhard Neidecker-Lutz** is the CTO of SAP Research, SAP AG's worldwide research organization since 2003. He joined SAP in 1999.

From 1988 to 1998 he worked at the European Applied Research Center of Digital Equipment Corporation in Karlsruhe, Germany. Mr. Neidecker-Lutz is a professional member of the ACM and holds a masters degree in computer science from the University of Karlsruhe.

be specialized protocols, because you're always operating at the limits of what's physically needed, physically possible."

Mr. Neidecker-Lutz predicts there will always be a "zoo of communication"—strange link layers, routing protocols, and so on. Many of the scenarios inevitably will be mobile, so IP won't always make sense.

"When people lament the lack of standardization, they are barking up the wrong tree," he says. "There is a set of different standards that are emerging in their domains, so you will have bundles of useful standards in various areas. This is not very well understood by some of the big players that want to shoehorn one-size-fits-all."

Mr. Neidecker-Lutz does not believe standardization will be as much of a barrier as feared. He expects RFID to be more limited by the cost of putting tags on devices than by the lack of standardization. As new technologies allow us to tag devices at lower cost, IOT technologies like RFID will proliferate.

"I don't think waiting for standards is going to solve anything," says Mr. Neidecker-Lutz. "There is a bundle of standards that you will need, depending on what you're trying to do, so I don't think, for the infrastructure itself, lack of standardization is the issue. You have to deal with the heterogeneity."

This puts a premium on the agile development of software and standards. In recent years, the agile software methodology has emphasized iterative development, adaptation, and speed. This reflects the move toward a more dynamic business environment in which companies increasingly operate in chaotic market conditions, improvise mashups, form ad hoc partnerships, and seize business opportunities. Standards should reflect this trend. No matter what form they ultimately take, the standards of the IOT should allow room for agility, ingenuity, and improvisation.

Indeed, standards are a technical matter, not a business matter. Standards must not be so restrictive that they impair later innovation. "Of course standards cannot really define the business," says Dr. Nochta. "The business is never standardized. There has to be room for creativity."

But there are limits to agile standardization when it comes to the IOT. Mr. Neidecker-Lutz warns that agile standardization runs up against the incumbency of existing systems. "The logistics of changing something in these embedded systems in an agile fashion is horrendous," he says. "You almost can't do it."

Oftentimes, says Mr. Neidecker-Lutz, this software is deployed in "a really ugly environment." There may be limited connectivity—or you may even be trying to change the connectivity itself. One example is software-defined radio (SDR), which uses software to replace features traditionally performed by hardware. SDR frequently has new variations but is curtailed by regulatory restrictions. Some of these regulations are archaic, notes Mr. Neidecker Lutz, and include some laws enacted in response to the sinking of the Titanic. In other words, wireless technologies of the IOT are being regulated by laws of the Marconi radio era. "We still suffer from some of the idiotic things that are now written into law in all the countries worldwide," he says. "Otherwise we could actually be a lot more agile about this stuff."

In other cases, the embedded systems may have built-in features that are essential to safety, such as those in cars or airplanes. Any updates might require requalification of the entire configuration. "So there is built in, sometimes for very good reason, resistance to being agile," says Mr. Neidecker-Lutz.

## Build It and They Will Come: The Power of Adoption

Market forces are likely to influence the adoption of standards. Recall Justin Rattner's assertion at the IRF that we should let Darwinian natural selection take its course. In other words, we should not be too

**Michael Bechauf** is Vice President of Industry Standards and the SAP Developer Network at SAP AG. He is responsible for managing SAP's participation in industry standards through a company-wide governance process that balances customer benefits with intellectual property risks. Mr. Bechauf serves in several industry activities, such as the Eclipse Foundation Board of Directors, as well as the Java Community Process Executive Committee. He is also managing the SAP Developer Network (SDN), an online collaboration platform for developers in the SAP ecosystem with more than 1.9 million members, as well as the SAP Mentor Initiative, a high-impact influencer group consisting of SDN developers, technologists, and bloggers. He has been with SAP since 1997 and lives in the San Francisco Bay Area.

heavy-handed about imposing standards and curtail creativity. Others take a similar view.

According to Michael Bechauf, Vice President of Industry Standards and the SAP Developer Network at SAP AG, standards boil down to two basic criteria: usefulness and market power. Is the product useful and associated with an application that people will find valuable? And does it have market power behind it? "In the end," he says, "it's about the power of adoption."

Mr. Bechauf puts little faith in a conceptual or academic approach to standards. He is skeptical of the idea that we can gather all the players in a smoke-filled room and come up with a top-to-bottom stack of standards for the IOT. "There is really no correlation between completed standard specifications and their adoption," he says. "It's two different things."

According to Mr. Bechauf, standards setting should begin with the application. The better approach is to build useful products. "What's the business problem that you're trying to solve?" asks Mr. Bechauf. "Prototype it and build software. When you want to drive adoption, then build a standard around it." Indeed, in most successful standards efforts, including the ones that are at the foundation of the Internet, working implementations preceded setting of standards.

Often, standards take root because of the dictates of big players with market power. Take the case of web services. According to Mr. Bechauf, they weren't adopted because they're optimal, but because Microsoft and IBM decreed "this is how it's going to be." Another example is EPCglobal. "The initial adoption wasn't really because the standards were perfect," he says. "The initial adoption driver was Wal-Mart and later regulatory forces in certain geographies."

A standard doesn't have to be perfect, adds Mr. Bechauf; it just has to be good enough. Standards should ensure that there are few barriers to adopting the product.

"You have to create something that has as few encumbrances as possible so it's easily adoptable and it's easily implementable," he says. "People will only engage in a forum that gives them an opportunity to build consensus and influence the outcome, as opposed to just ratify completed standards."

### Interoperability by Brute Force

Another perspective on standards comes from Tim O'Reilly and John Battelle, two insightful and well-respected observers of the digital world. In the paper, "Web Squared: Web 2.0 Five Years On," they argue that the standards barrier will be overcome by a combination of mass usage, machine learning, and semantic interoperability.

Mr. O'Reilly and Mr. Battelle point out that only 15 years ago email was fragmented into hundreds of incompatible systems joined by "fragile and congested gateways." Eventually, one of those systems, Internet RFC 822 email, became the standard. They expect a similar storyline for the next wave of standardization. As they write, "Vendors who are competing with a winner-takes-all mindset would be advised to join together to enable systems built from the best-of-breed data subsystems of cooperating companies."

The IOT is a central part of Mr. O'Reilly and Mr. Battelle's vision for the Future Web. Indeed, the term "Web squared" refers to the Web meeting the physical world. They believe this already is happening even as questions about standards remain unresolved. When most people think of the IOT, they tend to imagine conventional sensor technologies like RFID. Mr. O'Reilly and Mr. Battelle argue that the IOT already is seeping quietly into our everyday lives in the form of our smart phones. "But what many people fail to notice is how far along the sensor revolution already is," they write. "It's the hidden face of the mobile market, and its most explosive opportunity."

These phones contain cameras, microphones, GPS, and barcode scanners. All of these things are ways of sensing the physical world and rendering it in digital form. Most importantly, mobile applications are connected applications. These networks get better as more people use them and create a virtuous feedback loop that, in turn, inspires more usage. As these information shadows grow thicker, we can find more ways to link relevant data. As Mr. O'Reilly and Mr. Battelle write:

"Many who talk about the Internet of Things assume that what will get us there is the combination of ultra-cheap RFID and IP addresses for everyday objects. The assumption is that every object must have a unique identifier for the Internet of Things to work. What the Web 2.0 sensibility tells us is that we'll get to the Internet of Things via a hodge-podge of sensor data contributing, bottom-up, to machine-learning applications that gradually make more and more sense of the data that is handed to them. A bottle of wine on your supermarket shelf (or any other object) needn't have an RFID tag to join the Internet of Things, it simply needs you to take a picture of its label. Your mobile phone, image recognition, search, and the sentient web will do the rest. We don't have to wait until each item in the supermarket has a unique machine-readable ID. Instead, we can make do with barcodes, tags on photos, and other 'hacks' that are simply ways of brute-forcing identity out of reality."

## The Shadow of the Internet

Given the debate over standards for the IOT, it may be worthwhile to look at how standards developed in the past. The Internet is often cited as a model of standardization and interoperability.

The Internet was initially funded by the government and limited to research, education, and government uses. Commercial applications were rare until the 1990s. The Internet had its origins in the ARPANET of the 1960s. It gave rise to the communications protocol Transmission Control Protocol/Internet Protocol (TCP/IP) to facilitate communications between the growing number of networks. This became the standard for Internet communications. Yet many communications protocols existed, some proprietary and some competing. As one Internet pioneer, Ben Segal, recounted, "In the beginning was—chaos."[2]

This tumultuous state of affairs persisted for years. In 1989, Tim Berners-Lee and others at the European Laboratory for Particle Physics (CERN) proposed a new protocol for distributing their research. This became the hypertext transfer protocol for information distribution, which became the foundation for the World Wide Web.

Berners-Lee wrote hypertext transfer protocol (HTTP) and devised a way to identify unique document addresses, the URI or unique resource indicator. He added a browser named "WorldWideWeb" to view them and created the first web server to store and transmit them. Sound familiar?

Until the 1990s, the Web was in danger of becoming a mass of unrelated protocols. In 1994, Tim Berners-Lee and others founded the World Wide Web Consortium to promote standards for the Web, and this group has helped ensure that there are common standards for every browser. The result is the thriving Web we see today—one that is so successful that it is giving birth to new generations, such as the IOT.

A few lessons deserve to be drawn out. The standards of the Internet—a shining example of successful standardization—were laid down by people who emphasized basic functionality, not the parochial needs of one industry or one company. The standards were not drawn around proprietary technologies or specific products. The people who developed Internet standards were not the ones who ultimately profited from them.

The standards that laid the foundation of the modern Internet emerged only after a period of chaos. In this light, the present lack of clarity over standards for RFID and other IOT technologies seems less alarming and perhaps part of the natural progression of technology.

But people sometimes forget this history. Instead, they call for a top-to-bottom stack to be constructed up front. Dan Woods, moderator of the

---

[2]  Segal, Ben, "A Short History of Internet Protocols at CERN," *http://ben.home. cern.ch/ben/TCPHIST.html*

IRF 2009, suggested that this unrealistic expectation may be curtailing the IOT. "The success of the Internet makes people forget that nobody had this grand vision in mind at the beginning," Mr. Woods said. "But IOT standards-setting efforts often attempt to build a stack with massive scope. Perhaps in many cases the eyes of the standards setters are bigger than their stomachs."

Dr. Nochta echoes this point. When it comes to setting standards, the quest for perfection is getting in the way of progress. Even a mediocre standard is better than nothing.

"If everybody agrees on the same non-optimal thing," says Dr. Nochta, "at the end it's the optimal thing.

"I'm not saying we will find the technically optimal solution because we set standards. Far from it," says Dr. Nochta. "We need a standard in order to reduce implementation work for meaningful business processes and other applications based on the Internet of Things. If that standard is not perfect, then it's a non-perfect standard. But it's a standard."

Data Management

7

Perhaps the Internet of Things is incorrectly named. The value of the IOT comes from not the devices but the data collected by them. Indeed, a more appropriate name might be the "Internet of Data."

Instrumentation and standards are just first steps. Of more profound effect is the information produced by IOT. When we look closely at use cases, we see that the true value comes from seeing things with a better microscope, one that can look deeper than previously possible to find trends, dangers, hidden costs, and opportunities. Oftentimes, the data shows us things that we do not immediately recognize because they are so unfamiliar.

Data management presents many new challenges. How can we build new platforms to manage and analyze this information? How can we deal with the flood of inputs from all these sensors? How can this raw data be aggregated, distilled, analyzed, and synthesized into dashboards that provide useful information at a glance? How will important events be recognized and white noise be filtered out? How will these new schemes alter the competitive landscape?

## It's the Data!

The major challenges of the IOT are not necessarily technological ones. The larger question is how to handle the volume of new information that will stream forth from these devices.

"We get all hung up on the sensors and the protocols and the microcontrollers," says Mary Murphy-Hoye, Senior Principal Engineer in Intel's Embedded & Communications Group. "We're so looking at our hand right in front of our face that we don't look beyond. Most of the conversations I've heard until now have been about the instrumentation. I've been the one standing on a hill saying, 'But it's the data!'"

Ms. Murphy-Hoye recalled attending a recent conference with leading researchers in wireless sensor technology where many people were preoccupied with mundane technological matters like writing routing protocols, power efficiency, and calibrating nodes. But they lacked a broader perspective.

To put it simply: It's the data, stupid.

It's the data, says Ms. Murphy-Hoye, that allows us to create new business models. It's the data that allows us to see things that were invisible before. It's the data that will confer competitive advantage and separate winners from losers.

Only recently have logistics and supply chain managers begun to realize that data is the key innovation of the IOT. Devices are just hardware, but information is intelligence.

"We get hung up on how you do the instrumentation," says Ms. Murphy-Hoye. "We forget that this instrumentation creates data that no one's ever seen before."

Once we set up these systems, we have to learn to read the data—and that means learning a new language. This language is very different from what we have seen before. As Ms. Murphy-Hoye says, "There are characteristics about the data that we don't understand yet and, as a result,

**Mary Murphy-Hoye,** a Senior Principal Engineer in Intel's Embedded & Communications Group, is a supply chain and information technology solution development expert as well as a trusted advisor across numerous industries—end-to-end retail, industrial automation, high tech, oil & gas, chemical, aerospace and automotive manufacturing, and transportation.

Ms. Murphy-Hoye's recent focus has been the creation of Intel's RFID/Wireless Sensor Networks Lab for industry-scale proactive computing experimentation across businesses. She is currently developing multiple embedded Atom research pilots, working on instrumenting the US rail industry, and creating composable collaborative workspaces with Steelcase, a global office furniture leader.

Ms. Murphy-Hoye joined Intel in 1987. As a past Director of IT Research, she formed Intel's IT Research Agenda and specialized in research of disruptive technologies. During these twenty-two years, she has continuously reinvented herself by learning new technologies and gaining a broad understanding of Intel's business, to ensure better engagement within and across Intel groups.

aren't creating the capabilities that allow the use of it or the necessary insight to harvest its value."

In order to comprehend this data, we have to understand the physical context of the devices. We must take into account dynamic, spatial, and temporal data. We must think about these objects in the context of many other objects and about the proximity of those objects to each other and how that affects the information you get about them.

"There's this ebb and flow," says Ms. Murphy-Hoye. "I always describe the data using musical terminology because there's this cadence to it, there's also this movement of the object and the information about it. Those characteristics are very different and we need to start thinking about how you represent that."

This raises a host of new questions. How to manage this data? What timeframes are appropriate? How long do we keep this information?

How do we filter it? How do we design systems that sift through all the chaff and call attention to important events?

"There are all these questions that are very specifically application- and infrastructure-related that we haven't even begun to think about, much less solve," says Ms. Murphy-Hoye. "And this data is like a tsunami. How are we going to create something valuable from it?"

In many cases, we have no idea what our data is telling us at first because we do not know what questions to ask. Ms. Murphy-Hoye's reference to music is more than just a poetic analogy: in one case she and her colleagues literally turned the data into music in an attempt to make sense of it.

"What you tend to do at first, when patterns like this are emerging, is that you discount them, because it's not what you expect—'I've never seen anything like this so it must just be an anomaly with the nodes,'" she says. "You have to be able to step back and say, 'okay, what's happening is not what I expect; what does this mean?' There's a whole discovery aspect."

Again: It's the data, stupid.

> *"If we're the ones who figure out the data, then we win."*
> *— Ms. Murphy-Hoye*

"This is where all the revenue is, this is where all the value is, this is the differentiator," says Ms. Murphy-Hoye. "And if we're the ones who figure out the data, then we win."

## New Architectures

How do we handle all this new data?

What kinds of architectures are required? How should we determine the appropriate level of complexity and abstraction at each level? Say a railroad company puts sensors on all its rail cars. A manager might not need up-to-the-minute data about the temperature fluctuation inside every single railcar as it moves inch by inch through the system. But the company does need to aggregate data to schedule preventive maintenance or optimize its train schedule.

The IOT brings a deluge of new data. Consider the example of how smart meters might transform the information systems for electrical utilities. Today utilities take meter readings once a month. With smart meters, they would take readings every 15 minutes. Suddenly they would be dealing with a 300-fold increase in meter data, compressed timeframes, and many additional challenges. This new flood of data would oblige utilities to install big servers and analytical capabilities.

Therein lies the rub: the sum of data is increasing but the ratio of what is known to what is knowable is shrinking. We need technologies to sift through this information and make sense of it. Otherwise we'll be blinded by a snowstorm of information. Data analysis, visualization, machine learning for filtering data, and the ability to see patterns will become increasingly valuable skills.

Once this data comes streaming in, what do we do with it? How do we turn this flow into useful information? This is a major challenge of complex events processing. You may have data streaming in from thousands of locations, but all you really need to know is whether your shipment will arrive on time. This data needs to be aggregated and analyzed. Only then can it serve as the basis of intelligent decision-making.

"There is a big question about whether the system architectures that have been proposed can handle the billions and billions of queries that a successful Internet of Things will trigger," says Prof. Günther of Humboldt University, "If you imagine every object on this planet being tagged and sending out [location information], and you have literally billions of objects moving around, which infrastructure could mirror this real world in the virtual world and still provide efficient processing? This is a big challenge.

"If you start managing every screw in your enterprise, you're going to have a lot of data," says Prof. Günther. "You need some kind of hierarchical architecture to handle all of this."

According to Mr. Neidecker-Lutz of SAP Research the telecommunications companies may have some things to teach us about the IOT. The telecoms' experience with GSM, international roaming, dealing with a multiplicity of standards, and handling infrastructure makes them poised to play in the IOT. "If you look at the kind of infrastructure that is needed to do that, you realize that somebody's already doing this at pretty interesting scale, with the things being mobile phones," he says. "And if you look at their event processing, routing, billing, infrastructure, and so on, you get an inkling that somebody knows how to do this at scale."

*"The telecoms don't realize that they have most of this. I don't think they have the expertise to know what they have."*

*— Mr. Neidecker-Lutz*

But thus far this potential is latent, he adds. "The telecoms don't realize that they have most of this," says Mr. Neidecker-Lutz. "I don't think they have the expertise to know that what they have, if transformed properly, actually gets them into such a position."

## Central Platforms

We can think of the IOT as a series of sensory nodes, like the nerve endings that bring us data about what is happening in the world. These devices will have limited memory and power capability. Like nerve endings, they are useless unless they can feed into a central nervous system. Similarly, the IOT requires a system to manage all this data.

But how to filter and manage this data? Some data is useful only in the short term. But some data should be kept indefinitely because it can reveal important trends. Dr. Kubach of SAP Research makes an analogy to human short- and long-term memory. How can we create systems that filter and store information worth saving? And how can these systems incorporate various sources of data, from enterprise systems to customer forums?

We need semantic technologies to integrate this information. For example, back-end enterprise systems and a customer forum may use very different languages to convey information about the same product. SAP is part of a consortium of companies and researchers that seeks to tackle this problem in the Aletheia project. The Aletheia project (named after the Greek goddess of truth) is developing semantic technologies to harmonize product-related information from a variety of sources, such as ERP systems, emails, or Web 2.0 resources like blogs, wikis, or customer forums, and IOT technologies like RFID. Semantic technologies can help deduce meaning and thus make relevant information from many sources and many standards available to the user—something that existing technologies cannot accomplish.

"We can use semantic technologies to harmonize and to find out that those two people are talking about the same thing, the same problem, or the same feature of the product," says Dr. Kubach. "But how can we organize and bring this information together so that you can really make decisions based on this information?"

Dr. Kubach said enterprises need some kind of platform that integrates all these elements. They need a platform adaptable to different types of standards, data formats, software, devices, and Web 2.0 elements like customer forums. They also need better semantic technologies to understand all this information. In the future, IT systems will increasingly have to incorporate an array of technologies, such as auto-ID, automated meters, or factory machinery.

"It's all about device connectivity, more or less," Dr. Kubach says, "We should have a generic platform for this that fits all these needs."

Rather than trying to standardize everything—a task akin to herding cats—the key is to build a repurposable platform and then make connectors when necessary. Essentially, this is a matter of mapping one data format onto another. For example, it might have a connector that communicates with an OPC device and translates the data into XML format for the internal system. The connector approach is a proven model in software integration.

## Distributed Approaches

Some have suggested more distributed ways of processing this information, such as peer-to-peer approaches.

One option is putting more intelligence into the devices themselves. For example, smart electrical meters might be programmed to do data processing at the local level and share only summary data with the central system. Dr. Gregor Hackenbroich, Research Program Manager at SAP Research responsible for the area Data Management and Analytics, describes this as a "small data warehouse" that is linked back to the enterprise.

Similarly, Ms. Murphy-Hoye described how the devices themselves could help reduce the strain on the central systems. Ms. Murphy-Hoye has instrumented railcars in a number of studies. Putting motion sensors and accelerometers on railcars can help identify which cars show the distinctive vibration patterns that warn that wheel bearings need replacing. Preventive maintenance thus can reduce the risk of derailment. Proximity detectors could help warn when trains are in danger of collision. Instrumenting railcars could help railroads see the exact order of cars in a long train and tell exactly where each cargo container is in a long train that may contain hundreds of them.

"They don't have the infrastructure, or the money to create an infrastructure, that would tell them where the cars are all of the time," she says. "So they want to use the cars to give them insight as well as to help them understand what's going on with the cars themselves. That transforms their dispatching and scheduling, and it transforms their ability to offer additional services to their customers. They see it as a real differentiator." This is high-resolution management in practice, a topic discussed further in Chapter 9.

In short, putting more intelligence and processing capability in the nodes can help filter the data and reduce the strain on the enterprise systems.

### New Platforms, New Agility

Mr. Bullotta of Burning Sky Software is bullish about how the IOT may drive new platforms that enhance collaboration between people, devices, and systems. The platforms of the IOT will have real-world awareness and bring new visibility to enterprise systems.

Mr. Bullotta believes these changes are part of a larger transition from the industrial age to the information age. He cites the work of the famous futurist Alvin Toffler, who described these changes as a series of "waves." The first wave represented a distributed agrarian society made up of people who both produced and consumed their products, labeled "prosumers." The second wave was the shift to an industrial society, a more centralized structure that divided producers and consumers. Now we are moving into an information and knowledge-driven society and once again returning to a more distributed structure where information is distributed among "prosumers." (One caveat Mr. Bullotta notes is that we are not yet fully in the information age, only in the transition stage.)

This sets in motion a series of power shifts. Power becomes less about control of resources and violence and more about knowledge. Mass media, mass marketing, and mass manufacturing give way to more distributed forms—a trend dubbed "demassification." Social and work lives become less centralized. Organizational structures move from bureaucracy to "ad hocracy."

As a result, work will demand more temporary skill sets. Mass production and formal processes will remain, but will shrink to a smaller part of wealth creation. Ubiquitous computing and connectivity in devices and "things" will provide another powerful transformational capability.

Mr. Bullotta envisions "disposable companies." The founder might serve as the brand owner and recruit players for all the necessary processes: a design firm, manufacturer, marketer, distributor, and so on.

"All the pieces of that chain are getting broken apart," he says. "That creates a whole new dynamic that's needed for real-time communication and coordination of the information from that product and about that product."

Until now, we have applied IT to repeatable processes. The idea was that certain processes are done again and again and that IT could help create automated systems in an assembly line approach.

"The whole industrial mindset was: build a factory, repeatability, productivity, make the same thing a zillion times," says Mr. Bullotta. "That's all well and good, but there's a whole long tail opportunity of 60 or 70% of business processes that don't fit and haven't had IT applied to them very effectively."

Gartner recently estimated that 60% of business processes fall into this unstructured category. They're ad hoc, transient processes that meet a market demand, a last-minute customer order, or a disaster. As Mr. Bullotta says, "There's no flow chart, no process in your ERP system to deal with those situations."

How do we create platforms that allow us to seize these opportunities? This, says Mr. Bullotta, is where the IOT can add value to the enterprise. We need platforms that enhance business processes with real-world awareness. We need platforms that capture the knowledge gained in these processes.

"You may never see the exact process again," he says. "But by capturing that knowledge, it becomes very useful the next time. It's a searchable, reusable asset."

According to Mr. Bullotta, this represents the next wave in the evolution of enterprise software. This evolution began with monolithic applications and moved to composite apps, mashups, and more flexible business processes.

Companies will take advantage of the IOT, along with other tools like cloud computing, collaborations, mashups, Web 2.0, and social networking. They will outsource repetitive processes. This flexibility will allow companies to exploit the opportunities of the long tail and capture more business.

## New Fields of Battle

The IOT opens the door to new competition. The IOT creates new sources of data and this immediately raises questions about who controls this information. In building these platforms, players at each level will struggle for dominance. These struggles will occur at several levels.

## Data

In some cases, one party may control access to the data itself. For example, Wal-Mart controls all its sales data and uses that as leverage when bargaining with its suppliers. Data is power. The more one party can control this data, the more powerful they are. In some cases, the owners of the data may charge for access or for subscriptions or may use it for competitive advantage.

## Applications

In other cases, data may be freely available and the competition may occur at the application level. One example is flight tracking services. Flight telemetry data is made publicly available by the government and several companies offer web sites where users can check the status of flights. In this case, the fight for dominance occurs at the application layer. The competition boils down to the question of who can make best use of this data.

## PLATFORM COMPETITION: FLIGHT STATUS TOOLS

The Federal Aviation Administration (FAA) makes flight data available to the aviation industry. This data feed, known as the Aircraft Situation Display to Industry (ASDI), includes near real-time location of aircraft, flight plans, airspeed, and altitude.

Several companies provide flight tracking services for free or subscription. Since the FAA provides the data to all these companies, the competition occurs on the platform level. Competitive factors include proprietary algorithms, data management, data integration, usability and complementary services. Here are some of the players in this field in the US:

FlightAware **www.flightaware.com**

*Houston, Texas*

FlightAware was the first company to offer free flight tracking services to both private and commercial customers. FlightAware also provides commercial products and services. Customers include Continental Airlines, Cessna, Netjets, and Lufthansa.

FLIGHTSTATS **http://www.flightstats.com**

*Portland, Oregon*

FlightStats delivers real-time and historical flight information. Conducive Technology Corp., a private firm that provides flight information solutions, operates this platform. The company descends from earlier ventures that built booking engines for American Airlines, Air Canada, and Aer Lingus. FlightStats also offers a variety of business services to the travel and transportation industry (travel agencies, airlines, airports, publishers, consultants, airfreight government organizations, for example).

FLYTECOMM **http://www.flytecomm.com/**

*Mountain View, California*

FlyteComm was founded by former FAA employees who helped develop the government's Traffic Management System. In 1996, FlyteComm introduced WebTrax, the first web-based flight tracking system. Today its patented technology solves problems associated with tracking planes, weather, and flight logistics. FlyteComm develops real-time flight intelligence solutions for transportation, aviation, corporate travel, government agencies, and media companies.

*(continued on next page)*

*(continued from previous page)*

**FlightView®** http://www.flightview.com/

*Allston, Massachusetts*

FlightView provides a variety of products and data feeds. Customers include media companies, such as the *New York Times,* FedEx, Hertz, JetBlue, and United Airlines and major airports. Founded in 1981, FlightView was the first non-airline organization approved to receive Aircraft Situation Display (ASD) data from the FAA and developed the first commercially-available radar-based flight tracking system. In 1996, FlightView launched a web flight tracking site. FlightView is a brand division of RLM Software.

**flightwise** (formerly fboweb.com) http://flightwise.com/

*Orlando, Florida*

Flightwise (formerly fboweb.com) offers services to aviation professionals, enthusiasts, and the public. It makes flight data available to programmers and developers through its PlaneXML Flight Data API. Flightwise provides an array of services for flight tracking, information for pilot flight planning, and even a tool for scheduling and managing a fleet.

## Middleware

Another struggle will occur at the platform layer. Who can consolidate all this data and applications of the IOT into integrated platforms? One example is GPS systems, which are combining mapping data, turn-by-turn directions, restaurant review sites, and the like.

These struggles may span multiple layers. Rules at one layer may profoundly affect what happens at another layer. Consider the effect of open source. A lot of virtualization technology wouldn't work without open source. If Linux were not free, creating a new virtual machine would not be free either.

For example, the energy field may be heading toward a major shakeup. The smart metering domain has a great deal of heterogeneity. There are many players: device manufacturers, middleware companies, and upper-level companies like SAP and Oracle that provide business

**Dr. Gregor Hackenbroich** is a Research Program Manager in Data Management and Analytics at SAP Research. His main interests include the management of structured and unstructured data, data integration, and business intelligence. Dr. Hackenbroich and his team developed the Auto-Mapping Core that has been productized in SAP Business Process Management. He has been responsible for the acquisition and management of several large research projects, including the acquisition of the lighthouse project THESEUS. Prior to joining SAP, Dr. Hackenbroich spent two years at Yale University and was on the faculty of Essen University. He received his habilitation in theoretical physics from Essen University, and his doctoral degree in physics from the University of Munich. He has been a Heisenberg Fellow, and has won research fellowships from the Alexander-von-Humboldt Foundation and the Studienstiftung des Deutschen Volkes. Dr. Hackenbroich has published more than 50 research papers on computer science and theoretical physics, and regularly works as a reviewer for scientific journals and international research organizations.

applications. Collecting customer data every 15 minutes instead of reading meters once a month creates voluminous amounts of new data that opens the door for new players, new solutions, and new business models. According to Dr. Gregor Hackenbroich, similar battles are occurring in many other vertical markets and industries. The same dynamic is occurring with RFID, intelligent devices, manufacturing, and on many other fronts.

"This pattern repeats," Dr. Hackenbroich says. "What's true here in the Internet of Things is really this overwhelming amount of data, and that has the potential to completely shake up the field and bring in new players."

Mr. Neidecker-Lutz foresees battles on several levels.

The first round would be fought over virtual real estate—turf battles over who controls things such as domain registries and identity. Mr. Neidecker-Lutz believes that players like EPCglobal are essentially trying to turn virtual things into private property. "You have virtual real

**Virtual Real Estate**
Turf battles over control of elements like domain registries and identification. One example is EPCglobal in the electronic product code space

**Domain-Specific Operational Platforms and Central Infrastructures**
Competition by platforms to control specific markets, such as search engines or business services. One example is an anti-counterfeiting service

**Technology Infrastructure**
Battles to control the technology infrastructure for the Internet of Things. Potential players might include telecoms or giant companies like Google or Amazon

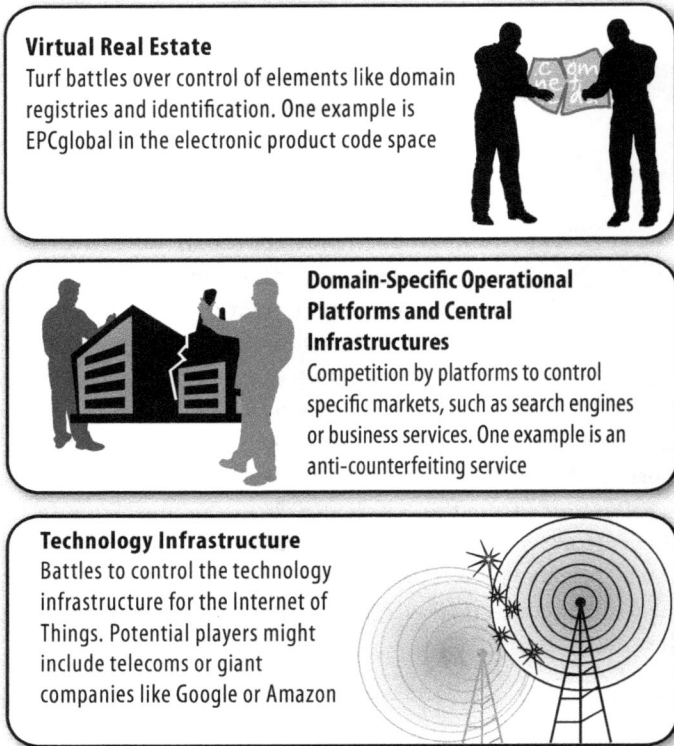

*Figure 7-1. IOT Battlegrounds*

estate wars at various layers, as with EPC. But since it is so fragmented, and many of the things in the Internet of Things already have identities of one kind or another, most of these games are not winnable for any monopolist…If you overstretch, people will figure out that they can just band together and provide an alternative."

Another interesting question is who owns the identification infrastructure. "It's less clear on whether or not that particular game can be dominated," said Mr. Neidecker-Lutz, "That largely depends on who is able and willing to operate the infrastructures."

Companies with operational platforms and central infrastructures for domain-specific functions will fight the second round of wars, he continued. One example is an anti-counterfeiting service, which sells not identifiers but authenticity.

In the area of infrastructure for search engines, databases, and event correlators, Mr. Neidecker-Lutz sees potential for the formation of natural monopolies, in which a single player, often an early entrant, comes to dominate because of economies of scale and barriers to entry. He predicted that "any functions that centralize a lot of data will be eyed very cautiously in this environment." The people concerned will vary case by case. With personal data, civil liberties groups like the ACLU may raise objections. With business data, the partners in the logistics chain may be wary.

Another battleground will occur in the technology infrastructure. This area is suitable to a cloud model because it is bursty and distributed. In theory, said Mr. Neidecker-Lutz, some of the metadata services like Google or Amazon should be capable of operating these infrastructures. "But for reasons I can't yet fathom they have so far chosen not to do that," he said. "It may just not have occurred to them."

The large telecom providers, especially those in mobile, also are potential players. "They're just using the technology for completely different purposes," he added, "but in principle the technology that they have in-house and know how to operate could provide much of the required functionality."

That said, much work remains to be done before any of these companies can make a big play in the IOT. Many applications, particularly those for complex event processing, simply don't exist. This is an active area of research in academia and at large companies like IBM. As Mr. Neidecker-Lutz concluded, "It's not entirely clear who will win that particular race."

Similarly, it is difficult to predict who will become the dominant platform providers. According to Mr. Neidecker-Lutz, IBM stands out as the only big company with the requisite expertise and technology in-house. The telecoms, on the other hand, have much of the technology, but not the required expertise. One possible scenario is that smaller companies team up with big telcos. Another possibility is that an entrepreneurial company makes a play on its own with an infrastructure provider like Amazon. Indeed, the cloud model reduces the barriers to entry and opens this field to many potential players. Only a few years ago the capital costs would have reduced the cast of potential players to a small number of big companies.

"That makes it very, very volatile, so I would not venture to guess how that's playing out," said Mr. Neidecker-Lutz. "In the past, it was fairly simple: it cost a couple of billion, so you only had to look at a few of the big players that could afford it. That's no longer the case."

## 8

In this chapter we survey some of the major barriers facing the Internet of Things. These barriers run the gamut: technology, business models, political economy, privacy, and psychology. The common theme is that each one represents an unresolved question regarding the IOT.

Here we touch on many themes briefly. Some major barriers like standards will be explored in greater depth in other chapters. We will briefly examine barriers in five main categories:

- Political economy barriers

- Technical barriers

- Business barriers

- Psychological barriers

- Regulatory barriers

It is important to point out that many of the topics covered here as barriers are only thought of as such because the ambition for the IOT is so vast. When the Internet was originally designed as a networking protocol, it would have been impossible to list all of the barriers we discuss in this chapter because nobody had the idea that it would change the world as it has and become a phenomenon that touches so many parts of society. But the IOT has no such innocence. The IOT is being created in the shadow of the success of the Internet and the implementers of the IOT are not trying to solve small problems but huge ones. These barriers, in other words, are barriers both to the next steps and to the achievement of the grandest vision possible.

## Political Economy Barriers

Standards and data management are often discussed as the major barriers for the IOT. But the IOT also raises difficult political and economic issues. How should the costs and benefits be divided? Simply put, who pays and who benefits? At times, these questions take us back to political theory. If these costs and benefits are divided unfairly, people may resist and slow the adoption of the IOT.

## The Supplier's Dilemma

Prof. Günther of Humboldt University believes that designing win-win partnerships is even more crucial to the IOT than standards. Indeed, the true power of the IOT comes from its network effects. The whole benefit of the technology unfolds when applied broadly across companies. If potential partners are loathe to join, this potential is diminished. Therefore, we must design more equitable models that allow partners to see clear benefits and encourage them to join. This is key to unlocking this power of network effects.

"These Internet of Things technologies are particularly tricky in the sense that costs and benefits are very unevenly distributed," said Prof. Günther. "Very often, the benefits accumulate at a few larger players, and the costs are, to a much greater degree, accumulating at some of the smaller players. That's why the smaller players say, 'why should we introduce technology if we have to pay and we don't get our return on investment?'"

This is precisely what happened with Wal-Mart, said Prof. Günther. The big retailer tried to order its suppliers to adopt these technologies with a heavy-handed mandate of "do this or you can't do business with us." Ultimately the suppliers balked and the whole system broke down. Unfortunately, he said, the industry failed to acknowledge this problem.

Metro took a more cooperative approach. According to Prof. Günther, Metro serves as a model for how to distribute the benefits more fairly. "They're basically doing cross payments to distribute cost and benefits fairly," he said. "And that is really the ticket."

Prof. Günther believes that designing win-win scenarios is even more crucial than the oft-mentioned issue of standards. Sharing can take many forms, such as cash, hardware, or information. The currency is negotiable as long as the costs and benefits are divided fairly among all parties.

*"Those who gain have to be willing to share."*
*— Prof. Günther*

As Prof. Günther said, "Those who gain have to be willing to share."

## Going Beyond Closed Loops

As noted repeatedly, the IOT remains largely confined to closed loops. Part of this is due to technological factors, the novelty of these devices, and the normal path of introducing new technologies. But it also reflects the problems of political economy. Deployments have been easier when the supply chain is controlled entirely by one company and there is no need to balance costs and benefits with partners.

One of the few industries where RFID has been deployed on a wide scale is in highly vertically integrated enterprises, like the clothing industry. Companies like the Gap essentially rule the whole supply chain, from production to retail store, and thus can see a clear ROI without the complications of involving other partners. How are we to expand these technologies more broadly among many partners?

"The holy grail is really using the Internet of Things in groups of cooperating companies, in complex supply chains," said Prof. Günther, "and that's difficult."

He compares this situation to the classic prisoner's dilemma. Cooperation between both parties would optimize benefits to everybody. But they resist because they fear that they will be stuck with the rotten end of the deal while their partners reap the benefits. This dilemma is often felt by first- and second-tier suppliers, who fear that they will pay the cost of RFID while most of the benefits go to the OEMs. Therefore these suppliers make a perfectly rational decision not to adopt RFID.

As with the classic prisoner's dilemma, the supplier's dilemma can be resolved with better communication. According to Prof. Günther, both sides need to come to the table, quantify costs and benefits, and divide them fairly.

Only then, said Prof. Günther, will cost-benefit considerations point companies toward adoption. At this point, most companies take a short-term, limited view of RFID and the IOT in general. They make operational use of RFID to make their processes faster, more efficient, or more secure. A more advanced and advantageous application would be to use RFID for strategic long-term purposes, such as better quality guarantees or other things that affect competitive standing.

Similarly, most RFID applications are intra-enterprise, closed loops where they can see a positive ROI in the near future. The more advanced situation that offers the largest systemic benefits is inter-enterprise chains involving other companies.

Prof. Günther illustrates these dynamics with a 2x2 matrix. Right now, most companies are in the lower left: intra-company, operational

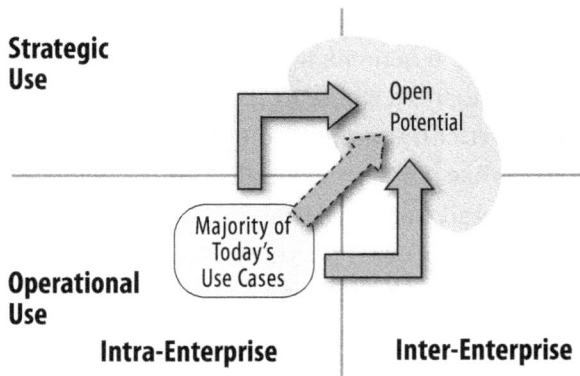

*Figure 8-1. The 2 x 2 Matrix*

applications that pay off in one or two years. "And where we really want to go is the upper right quadrant, with inter-company applications, really strategic uses of the technology, and new services for customers," he said. "That's the holy grail. That's where the sweet spot is, but it's difficult to get there."

### Sharing Benefits with Partners

How to find equitable sharing agreements? One solution is to look beyond the devices and ask what new value can be created for the various players.

Who pays and who benefits? Mr. Neidecker-Lutz of SAP Research says the typical model is when two independent pieces of business are combined a third thing is found to pay for it. This is the familiar concept of monetization. Take Google as an example. How does the search engine provide its services for free? By selling advertising. Similarly, Mr. Neidecker-Lutz expects that the IOT will be monetized, and costs amortized, by selling services enabled by the infrastructure. Devices will enable services, and services will pay for the devices. The Internet of Services and the IOT will have a symbiotic relationship.

This means that partners will have to find working combinations of infrastructure and services where each partner can extract value. As Mr. Neidecker-Lutz puts it, the key is "each of the players have enough of local value extraction from the combination of services running along this stuff to be able to sustain everybody's operation."

Partners should look beyond the instrumentation. What new data can be derived from the IOT and how can it allow partners to create new value or offer new services? How can it make them more efficient? Partners should not approach the IOT from a mindset of squabbling about pieces of a fixed pie. Instead, they should look at how the IOT can transform their businesses and create a bigger pie.

### Sharing Benefits with the Consumer

One way to create new value is by offering services to the consumer. Many versions of this are now available. Companies like UPS, FedEx, and DHL allow customers to track packages via the Web. Domino's even offers a "pizza tracker" web page where customers can check if their order is in the oven or in transit.

**Prof. Dr. Ryo Imura** is a founder, President, and CEO of Mu-Solutions company in Hitachi Ltd., which was established in 2001 to promote the world smallest RFID "Mu-chip." In 2004 he was an Executive Managing Director of Tracing & Tracking RFID Systems Division. Currently, Prof. Imura is Global Business Director in the Information & Telecommunication Systems Group and also Corporate Officer for Hitachi, Ltd. In addition, he is a professor at the University of Tokyo and a guest lecturer for MBA students at the Haas School of Business at the University of California at Berkeley.

The issue of sharing costs and benefits extends beyond our traditional business partners. Consumer acceptance will be a major hurdle for the IOT. The public may be suspicious of these technologies as they permeate ever more deeply into our everyday lives. If they can see clear benefits, they will be more likely to accept them.

According to Prof. Dr. Ryo Imura of the Mu-Solutions company in Hitachi Ltd., the biggest key to the IOT is showing clear benefits with practical applications. In the case of Wal-Mart, the mega retailer mandated the use of RFID tags, burdened suppliers with the cost of adoption, and didn't share any of the benefits.

"We need a clear mechanism for cost sharing," Prof. Imura said. "This is a political and economic issue that prevents cooperation and investment."

According to Prof. Imura, the network itself is the killer application of the IOT. Prof. Imura urges making this network as open as possible and even accessible to consumers. If people can reap advantages, they will be more welcoming toward the IOT. For example, an auto supply chain might be useful for a customer who wants information for car maintenance. Spreading the benefits widely will facilitate adoption.

Another example is food supply chains. As it now stands, customers can obtain relatively little information, such as name of manufacturer, price, and expiration date. In the future, they might be able to trace where farm ingredients came from or who made the final product.

"Food industries or food supply chains have very good services for tracing forward and tracing back if they have some access," said Prof. Imura. "But the customer or consumer would like to know what happened. The data management system is networked and it should be open, even for the customers."

The value of transparency goes well beyond logistics. For example, an organic food producer might use digital product memory as a form of customer relations. If a shopper checks a carton of organic milk and sees that it was produced on a certain date by a family farm that is certified organic, they may feel reassured and be more willing to pay a premium price for that product.

## Technical Barriers

Technical barriers include scalability, governance, reliability, device limitations, incumbency of the installed base, and cost of deployment. While these barriers are daunting, compared to the business and psychological barriers, they may turn out to be the easiest ones to overcome.

## Scalability

As noted in the previous chapters, the IOT raises architectural challenges, due to the volume of data. It remains to be seen whether system architectures will be able to handle the billions of queries from the IOT. Some have suggested more distributed ways of processing this information, such as peer-to-peer approaches.

"Primary scalability issues are not solved," said Prof. Mühlhäuser of Technische Universität Darmstadt.

If we tag every retail item and track it anywhere in the world, we'll generate vast amounts of data. "The systems we're designing currently will not scale up to a global scale," Prof. Mühlhäuser said. "The good news is that we've still got a couple of years until we'll get there. The bad news is that we are not investing enough energy on this issue."

The IOT will generate vast amounts of data. "The volumes are very small, but the transactions are unbound," said Prof. Mühlhäuser. Until now, data entered by humans has been limited by the size of our population and the cognitive limits of our processing ability. The IOT potentially removes these limits and turns the flow of data into something like a bot attack. As Prof. Mühlhäuser observed, "With machine-to-machine communication, there's no restriction at all."

## Governance

The IOT lacks a clear model for governance. As we saw in the discussion of standards, there is no universal numbering scheme. What should be the governance model for the IOT? Who controls the numbering scheme for identification? Where is all this information stored? The EU has proposed a more distributed international governance scheme because they fear domination by US companies.

This raises political questions. For example, EPCglobal has outsourced Object Naming Service (ONS) to VeriSign, an American company. VeriSign also manages the Domain Name System (DNS) directory for .com and .net. "Europeans don't like VeriSign because they don't like the idea that it is not only the Internet, but now also the IOT that is dominated by some US private vendor structures," said Prof. Günther. "I don't think the other countries will accept any kind of US-led standardization effort. I'm certainly not anti-US, but that's my feeling from the people I talk to in the political arena."

According to Marisa Jimenez, Public Policy Director Europe of GS1, EPCglobal contracted VeriSign to manage its ONS. But VeriSign does not manage any numbers. Under this distributed model, the information remains under local control. VeriSign does not manage any information about products, only the addresses of companies in the ONS directory. As Ms. Jimenez explains, "In case someone wants to know where to ask for information about a product, they can ask the ONS to get a point of contact for the company that owns that EPC manager number."

## Reliability

More sensors mean more physical exposure. These sensors are, by definition, open to the physical world. They are subject to direct physical assault or denial of service attacks.

Many of the existing networks lack sensor management capability. This challenge is all the trickier because of the limited power and processing available on edge devices. "That means also we have to investigate self-organizing schemes for balancing the reliability and

**Dr. Volkmar Lotz** is the Research Program Manager for Security and Trust at SAP Research. His responsibilities include the definition and implementation of SAP's security research agenda, its strategic alignment to SAP's business needs, and the maintenance of a global research partner network. His group's research topics include service security, business process security, secure software engineering, trusted collaboration, compliance enforcement, and adaptive security. Before joining SAP, he headed the Formal Methods in Security Analysis group at Siemens Corporate Technology, focusing on security requirements engineering, evaluation and certification, cryptographic protocol verification, and mobile code security. Dr. Lotz has been the main contributor to the LKW model, a formal security model for smartcard processors. In addition, his experience includes context-aware mobile systems, legally binding agent transactions, and authorization and delegation in mobile code systems.

the trustworthiness of information provided," said Dr. Volkmar Lotz, SAP Research.

On one hand, there is a desire to increase the reliability of these networks by deploying them widely, taking frequent readings, and making the nodes sensitive enough to detect events. On the other hand, this greatly increases the volume of information and raises the risk of false readings. As Prof. Österle of the University of St. Gallen showed in his IRF talk, making his house alarm system work was surprisingly tricky. Is the sensor really indicating an intruder in your shipping yard? Or is it just blowing leaves?

Dr. Lotz and his colleagues already have begun to propose potential solutions. "We address this problem on the middleware layer," he said. "We allow the applications to securely consume information from the sensor networks, even if the sensor networks themselves might not be completely reliable and trustworthy."

According to Dr. Lotz, they are examining secure middleware where "you collect the information but at the same time apply schemes that allow you to increase the reliability and trustworthiness of the information—for instance, by introducing redundancy, observing events, classifying critical events, reacting to critical events, reconfiguring your sensor network, and introducing additional nodes."

## Device Limitations

The functionality of devices is limited by battery power. These devices must be small, durable and, in some cases, capable of running on battery power for years. This also limits their storage and information processing capabilities. The IOT needs new advances in small battery power, energy harvesting, energy storage, and energy consumption.

## Incumbency of the Installed Base

The installed base also becomes an obstacle to innovation. Technologies in the field can become obstacles to innovation of new systems. In some cases, there is more freedom when engineering an entirely new system. This is similar to the limits of agile computing: the installed base becomes a barrier to innovation.

Likewise, one major barrier is the success of barcodes, which have been adopted around the world in manufacturing, retail, and many sectors. The universal acceptance and widespread use of barcodes may make some users reluctant to abandon it for a newer technology like RFID.

## Cost of Deployment

Cost of tags is a major issue. In general, tags are treated as disposable and it is impractical to reuse them. Fortunately, the trends are toward lower costs. For example, Prof. Günther's group did a study of a large clothing company and found that it became economical to deploy tags if they cost 6.5 cents or less.

## Business Barriers

Business barriers include the lack of a clear business case, the dearth of proven business models, questions on return on investments, payment structures, the expansion into closed loops, the lack of consumer applications, and questions of security and trust. These problems may seem intractable at first, but often vanish once a killer application demonstrates tangible value.

## The Unknown Business Case

One barrier is the unknown: how can you measure the benefits of something you can't yet see? This is akin to trying to justify a diagnostic procedure like an X-ray. According to Prof. Fleisch, ETH Zürich and University of St. Gallen, companies need to take it as an article of faith that these systems will bring insights that ultimately justify their cost. Early adopters, added Prof. Fleisch, tend to be strong-minded CEOs from midsize companies who learned from history that they can outperform others by early adoption of a technology.

"We're still at the very beginning of this curve," said Prof. Fleisch. "You need leaders who strongly believe in something because you can't calculate the benefits."

## Lack of Business Models

Much of the discussion of the IOT focuses on technical aspects. But Prof. Österle believes the more fundamental basic problem is a lack of understanding about how to create consumer processes and business models.

These models are complicated by several factors: How should we share development costs? How should we divide revenues? Who should orchestrate these systems? And how can we achieve the market size to make them viable?

This harkens back to the experience of Prof. Österle when he wired his house with sensors. The problem wasn't the sensors *per se*, but the systems that were supposed to knit them all together.

"These processes are still much too complicated," said Prof. Österle. "We don't understand the processes enough so that the technology, hardware, and software are really able to provide the services in a way that we need it. It's simply in an experimental state."

We will explore business models in greater detail in Chapter 10—and show how this barrier is being overcome with the emergence of new models.

## Return on Investment

Many companies are wary of the IOT because they cannot see a clear return on investment. The initial investment for hardware and software may be substantial. According to Prof. Österle, the return on investment may seem small because revenues are divided among so many partners. Often there is a long period with little revenue, just as we saw in the early days of Internet startups, many of which went under. Even Amazon and eBay took years to become profitable.

"It is about very small revenue pieces," says Prof. Österle. "The consumer uses 20 services a day, and each of them a very small amount, and it is hard for these companies to approach a big enough consumer or customer segment to really make sense for them in business terms, in revenue terms."

## Payment Structures

Another challenge is payment structures. If we live in a world where things constantly talk to each other and exchange information, how do we define new models for payment? The current web-shopping model, where the person pays with a credit card every time he or she encounters an item, simply won't scale to the IOT. There will be too many transactions for the current model to work when machines are talking to each other constantly. We need a model with automatic micropayments for a constant stream of small transactions.

## Closed Loops

The IOT is a vision with global aspirations; as a reality, it remains strictly local. Most successful examples represent closed loop applications. The challenge is to expand the closed-loop applications into open-loop scenarios.

This has proven slower than hoped. Dr. Christian Floerkemeier of the Auto-ID Labs at the Massachusetts Institute of Technology recalled with a chuckle that the MIT Auto-ID labs once predicted that we'd soon be able to track every soda can around the globe.

"Why didn't it happen?" he asked. "Well, the organizational complexity of implementing, such as a track and trace system on a global scale, is huge. The lesson learned from the Wal-Mart rollout is that just asking your hundreds of top suppliers to tag goods gains you little if the remaining suppliers do not follow, because, at the end of the day, change comes from analyzing data and doing business process reengineering.

**Dr. Christian Floerkemeier** received his Bachelor and Master of Engineering degrees in electrical and information science from Cambridge University in 1999 and his Ph.D. from ETH Zurich, Switzerland, in computer science in 2006. He is currently Associate Director of the Auto-ID Lab at the Massachusetts Institute of Technology.

Before joining the Auto-ID Lab at MIT, Dr. Floerkemeier was Associate Director of the Swiss Auto-ID Lab at ETH Zurich. From 1999 until 2001, he worked as head of software development for Ubiworks, an Amsterdam-based software company. Dr. Floerkemeier is the co-founder of the leading open source RFID project Fosstrak and is a member of the EPCglobal Architecture Review Committee. He was the Technical Program Chair of the first international Internet of Things Conference and IEEE RFID 2009, and has published numerous papers on radio frequency identification and pervasive computing.

If you only have 30 or 40% of your items in a certain business segment tagged, you can't do that."

At present, the successful applications are local. But Dr. Floerkemeier doesn't necessarily see this as a bad sign; rather it's a typical of an early stage technology.

Another reason that IOT applications have been most successful in closed loops is that the players can see clear benefits and return on their investment. This kind of scenario also avoids the nettlesome problems of sharing costs and benefits described in the political economy section. "The party that is benefiting from it is typically also the party paying for it," said Dr. Floerkemeier. "You don't have to do complex schemes where I'm willing to pay for my tags or my infrastructure, but I'm being paid back in terms of data access from my supply chain partners—these things are awfully complicated in practice."

## Lack of Consumer Applications

Thus far, most of the IOT uses are characterized by closed loops, B2B, and efforts to improve efficiency. Consumer applications are conspicuously absent from the IOT.

"If you look at Web 2.0 and the Internet, it wasn't driven by business," Dr. Floerkemeier said. "It was driven by consumers. The question is, how are we going to get that in the Internet of Things world?"

He believes that consumer applications remain a few years away. "One of the challenges is how to come up with real use cases that benefit the consumer," he said, "because then you have mass market potential and we're going to see probably the boom that we've seen around buzz words such as Web 2.0, and the Internet."

But there are a few early consumer applications emerging. Ford now offers an F150 truck with an RFID reader in the back that reads tags on DeWalt tools. The inside display lists all the tools in the back of the truck so that when contractors leave job sites, they no longer have to physically check whether their tools are packed away.

One benefit of a consumer-focused killer app is that it may demonstrate clear benefits and accelerate standards and governance.

## Security and Trust

Are businesses willing to freely exchange data with partners? As Dr. Floerkemeier says, "There are parties who might not be interested in sharing data because their business model relies on keeping certain data confidential."

How to develop mechanisms that allow companies to form IOT networks without compromising the security of their data? The data sits in different systems. "Technically bringing data together is one thing," said Mr. Neidecker-Lutz, "It can be difficult, but it can be done, depending on money," he said. "The more interesting thing is: can I do the queries I want without exposing too much of the innards?"

In logistics chains, participants may not want to share data. They will not be comfortable with a Google-like model where all data goes into a single place and can be accessed by anybody. The key is how to create an entity that everybody trusts—or even create a system with no need for such an entity. We need to adopt new forms of secure computing for the IOT.

According to Mr. Neidecker-Lutz, privacy-preserving computation has made some advances in these areas—but only to a point. Until now, these technologies have been limited to relatively smaller scales.

"There's a huge challenge of doing this at large scale with high data rates and high data volumes." He adds, "It remains to be seen whether that can be done."

Until now, many of these applications have been in closed environments where it is easier to exert control. As IOT applications spread to more open loops, this becomes more challenging. Enterprises may have to give up some control and accept whatever level of security the provider is willing to guarantee. Given that relations will be short and dynamic, it is not practical or feasible to renegotiate these arrangements in every case.

In fact, Dr. Lotz is not sure that complete security is attainable in the IOT.

"If we look at the Future Internet, which comprises the Internet of Things and the Internet of Services and all these largely distributed, billions of entities containing computing and application environments, I think we have to accept the fact that we will not be in a position to make this whole thing a secure and safe place," said Dr. Lotz. "So the

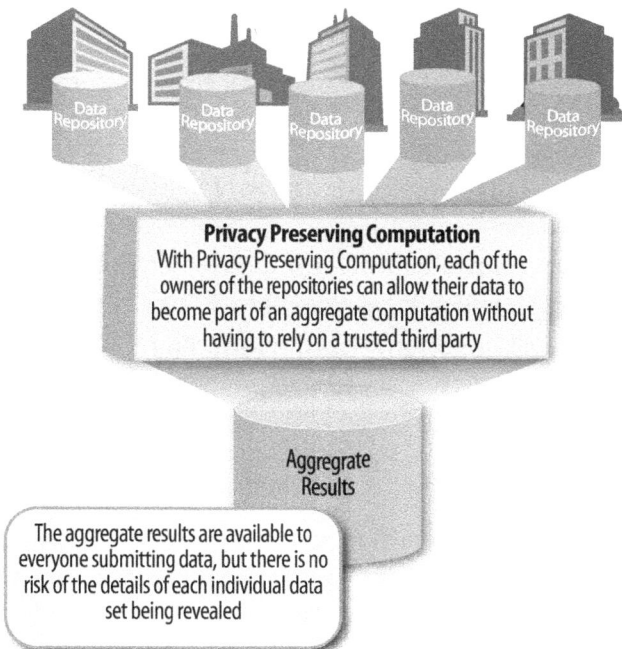

**Privacy Preserving Computation**
With Privacy Preserving Computation, each of the owners of the repositories can allow their data to become part of an aggregate computation without having to rely on a trusted third party

Aggregrate Results

The aggregate results are available to everyone submitting data, but there is no risk of the details of each individual data set being revealed

*Figure 8-2. Privacy Preserving Computation*

challenge is rather to make a safe place of those entities that you are currently working with and simply accept the fact that complete security for the whole Internet of Things is something that we will never be able to achieve."

## Psychological Barriers

The psychological barriers include skepticism, the fear of surveillance, and the problem of being blinded by the flood of data.

## Skepticism

Prof. Min of Fudan University believes the major barriers to the IOT are not technical but psychological.

Some people have overly optimistic expectations. According to Prof. Min, many companies eagerly plunged into the IOT and became frustrated by the slow return on investment. A second group was skeptical from the outset and remains unwilling to pursue the IOT because it remains unproven. Both views, he said, are major barriers.

"A wiser approach is to understand that the Internet of Things really is in its infancy," he said. "Infancy means we need to grow, and we need nutrition to grow."

We need to nurture the IOT. Key points: "First, start small," advised Prof. Min. "The second thing is to start with a project that adds value to the product." These small applications will help us understand how the IOT really works. Once these pilots prove themselves, the IOT will gain momentum and applications will become more complex.

We need to pick applications where the IOT can add real value. Cost reduction alone isn't likely to drive adoption, said Prof. Min. He works closely with supply chain management and logistics partners in China, and they want better visibility and value.

## Fear of Surveillance

The IOT raises many justifiable concerns about privacy. The perception of the danger lags behind the true reality of the threat. Dr. Lotz believes the IOT increases privacy concerns exponentially.

Yet many people remain ignorant of just how much these systems are capable of snooping into their lives. They remain unaware of how sensor networks can track their movements and pull up personal information. Moreover, technology allows this information to be distributed across the world instantly and used in many different ways.

Dr. Lotz believes that this naïveté will vanish once people realize just how much privacy they have lost. To prepare for this backlash, he suggests that industries should pursue "privacy-friendly" information and sensor networks. "People indeed do not care too much about the potential breach of privacy," he said. "However, that might change, and clearly the suppliers of this technology need to be aware of this issue."

Claudia Alsdorf, CEO of Original1, a joint venture between SAP, Nokia, and Giesecke & Devrient, agrees. Initially, she expects that consumers will be fascinated by the technology and enamored by all the new applications. Over time, they will become more guarded as they realize the potential for surveillance and just how much these technologies will reach into their lives. "I'm very sure that in the long run there's a kind of paradigm shift," she said. "The consumers will be way more cautious with regard to when and where they are using their smart phones and when they have the feeling that they can be even tracked by themselves."

On the manufacturing side, she expects the opposite: initial reluctance followed by zealous adoption.

In their research in the retail sector, Prof. Günther and his colleagues found that consumers do not resent electronic surveillance inside the store because they regard it as the domain of the company. But once the consumer steps outside, he added, they become much more protective of their privacy. "Once the good is sold, then the psychology changes," he said.

In one study, Prof. Günther and his colleagues offered consumers two choices. One option was to gain some of the RFID advantages post-sale while blocking personal information. The second option was to completely remove the tag. "I'm telling you, the vast majority wanted the thing to be cut off—even though the protection that we offered was as safe as a credit card," he said. "Not enough, they wanted the thing to be cut off. People are very sensitive post-sale."

Yet, Prof. Günther added, people might be willing to live with tags, and other IOT technologies, if they see clear advantages. After all, social networking has proven that people are willing to share all kinds of personal

**Claudia Alsdorf** has ten years of experience in executive management, development, licensing and commercialization of new consumer electronic products and services across the domains of the Internet, online commerce exchanges, virtual reality, and wireless communications. Five years ago, she founded Echtzeit AG, a global provider of 3D and online exchange products and services, and she continues to serve as the company's CEO. In 2002 she joined SAP as Vice President of Communications Development within Global Communications, a position that was responsible for communication strategy and long-term plan development, including aligning the communications strategy with non-corporate communication units. In this position, she spent one year in SAP's New York office, working together with the Global Marketing team. In 2004 Ms. Alsdorf became the Head of SAP Inspire, the internal venture group of SAP worldwide. From 2006 to 2009, she was also responsible for SAP Research Communications. In 2010 she became the CEO of Original1, a joint venture between SAP, Nokia, and Giesecke & Devrient, that delivers solutions that fight against counterfeits and product piracy.

information if they get something in return. Perhaps they would accept tagging if it gave them advantages like theft protection, instructions, or the right to return the product within a year with no questions asked.

Over time, Prof. Günther expects that these concerns will diminish. "If it's cheap enough, people will leave the tag on their milk carton or on their Corn Flakes box because they like the idea that their refrigerator tells them that they need to buy something," he said.

Other experts agree that people may be willing to trade privacy for value. Prof. Dr. Marc Langheinrich, Assistant Professor for Computer Science at the Università della Svizzera italiana (USI) in Lugano, Switzerland, researches security and privacy in ubiquitous and pervasive computing. He believes that the public will accept IOT technologies if it makes their lives easier or more enjoyable.

For example, in Switzerland people commonly use ski passes with embedded RFID tags because it allows them to glide quickly through the lift entrance. In fact, people even use RFID-enabled bankcards for the same thing. "If there's a value proposition, I think people are comfortable with disclosing personal information, given that there's something in it for them," he says.

But people do get upset when they discover that a company has used these technologies without their consent. "So I don't think there's a problem with RFID or any of these other smart technologies *per se*," he said. "You just have to create a value proposition for use."

We already are seeing early examples of these kinds of personalized value propositions. One famous example is the BMW Mini "Motorby" billboards that read ID information embedded in the drivers' RFID-enabled key fobs and can then display personalized messages.

Prof. Langheinrich believes that there will be no single solution to guarantee privacy. It is going to be addressed application by application, case by case. The solution will vary, depending on circumstances and stakeholders. (We will see examples of such variable solutions in the supply line section in Chapter 10.)

"What you really need is first to understand what the people are doing with the stuff you provide, what their interests are, and what they're willing to trade or disclose in return for what," Prof. Langheinrich

**Prof. Dr. Marc Langheinrich** is Assistant Professor for Computer Science at the Università della Svizzera italiana (USI) in Lugano, Switzerland. He received a master's degree in computer science from the University of Bielefeld, Germany, in 1997, and earned his Ph.D. from the ETH Zurich, Switzerland, in 2005. Before joining USI in September 2008, he worked as a researcher at the ETH Zurich, at NEC Research in Tokyo, Japan, and at the University of Washington in Seattle. Prof. Langheinrich has published extensively on security and privacy issues in ubiquitous computing, RFID, and the Internet of Things. He has been the program co-chair of both the 5th Intl. Conference on Pervasive Computing (Pervasive 2007) and the First Intl. Conference on the Internet of Things (IOT2008). Prof. Langheinrich is also the general co-chair of the 11th Intl. Conference on Ubiquitous Computing (Ubicomp 2010).

said. "Then what's the best security solution? It might not always be the strongest encryption available; it could also be something very simple, yet easier to use. We need more creative solutions, depending on the particular application scenarios."

## Blinded by Data

Another psychology question pertains to human limits. Are managers really ready for this flood of information? Or will it overtax our ability to absorb it all.

"How much information can a human being really bear before breaking down?" asked Ms. Alsdorf. "In psychology, there is the theory of curiosity. There's an optimal level of complexity, and, if you overdo it, you just say, 'no thank you.' If you do not have enough complexity, you are bored...It will be an art in management to filter the important information to make the right decisions in real time. I'm really not convinced at all that anybody is able to handle this complexity anymore. And the more data you put on managers, it gets even worse."

On the bright side, more voluminous data doesn't necessarily mean overload. Tim O'Reilly and John Battelle underscore this point in their paper, "Web Squared: Web 2.0 Five Years On," and cite Jeff Jonas' work on identity resolution. Mr. Jonas' database grew to about 630 million "identities" before the system had enough information to identify all the variations. At a certain point, his database began to learn and then to shrink through a process known as "context accumulation." Every object has an "information shadow" in the digital world: a book, for example, has a shadow in an ISBN, library catalog, Amazon, and so on. As Mr. O'Reilly and Mr. Battelle write, "As the information shadows become thicker, more substantial, the need for explicit metadata diminishes."

## Regulatory Barriers

Government authorities are certain to weigh in on the IOT. Public authorities are certain to take an interest in matters such as privacy, data security, and health.

The IOT will happen with or without government regulation. Yet these technologies reach into many aspects of our lives where personal privacy and public welfare are at stake. At the personal level, these technologies can literally keep us under surveillance. At a societal level, they can affect vital areas like food, drugs, environment, health, finances, or energy supply. These are all areas that already have been deemed worthy of government regulation and it seems inevitable that public authorities will have some say in how the IOT technologies are deployed.

As one European Commission (EC) report observed, "Simply leaving the development of the Internet of Things to the private sector, and possibly to other world regions, is not a sensible option in view of the deep societal changes that the Internet of Things will bring about."

Policymakers are likely to take an interest in several fronts. The IOT lacks a clear vision for foundational building blocks, such as standards, object naming, governance, and information security. Failure to protect these building blocks could stifle innovation and compromise privacy and security.

The EC has signaled that it intends to be an active participant in the IOT. For example, it has urged widespread adoption of Internet Protocol (IPv6) as the basis for the Future Internet.

The EC has highlighted key IOT topics of interest to governments:

- Governance

- Privacy and protection of personal data

- Trust, acceptance, and security

- Standardization

- Research and development

- Openness to innovation

- Institutional awareness

- International dialogue

- Recycling and waste management of devices

- Future developments

Prof. Mühlhäuser sees a potential role for government. He notes that in developing countries governments often take a lead role in building infrastructure. We are at a similar stage with the IOT and Internet of Services. "I tend to believe in the need of governmental influence on basic infrastructure," he said.

According to Prof. Mühlhäuser, the Internet of Services won't take off until governments and large alliances of corporations get behind it. He believes the Internet of Services, and possibly the IOT, requires attention and deliberate design from government and standards bodies to create platforms that are trusted and scalable.

"Now we are trying to transport our entire economy onto the Internet," he said. "Today and tomorrow, software—in the form of services—plays the role of commodities in this economy, often as a proxy for the real-world service it represents. But increasingly, it will also play the role of the dealers, of the agents in this economy, with the potential to transform its very nature. The change will be more dramatic than what we saw when software replaced agents in the stock market and recently in the financial market. These are reasons why politicians should have an incredible amount of interest in looking very carefully at what happens there and having a great deal of influence."

There is precedent for successful resolution of these matters. In Europe, government institutions and industries worked together to form the GSM8 standard, which became universally accepted.

Governments are likely to take an interest in addressing questions like network access, roaming, or billing. Government authorities in some countries are likely to be wary of any situations that may lead to monopoly or restrict fair competition. They may take steps to prevent private companies from restricting access to networks or controlling Object Name Server (ONS) management. Another area of concern is the reallocation of the radio spectrum with the proliferation of RFID.

Privacy is certain to be a subject of regulation. In fact, many analysts predict that the demand for privacy protection is likely to increase once the public realizes just how pervasive these technologies will become and how much they can snoop into their lives. For example, some privacy advocates have called for laws to guarantee the "right to silence of the chips"—or the ability of citizens to opt out of a network connection at any time.

Government mandates also may play a role in driving the IOT forward. This need not be a requirement to deploy certain technologies *per se*; but as governments demand more traceability for products, like pharmaceuticals or food, enterprises may see technologies like digital product memory as the best route to compliance. In the course of meeting these requirements, they may realize that they can expand upon these technologies for business reasons. In the US, mandates by Wal-Mart and the Department of Defense were major drivers for the adoption of RFID by suppliers. A similar phenomenon may occur with government mandates for regulation of consumer protection, health, or environment.

Dr. Zimmermann of TU Berlin points out that government mandates may also accelerate the adoption of the IOT. The mandates do not necessarily have to be direct. For example, the Dutch government liberalized electricity distribution and set more stringent requirements about the reliability of electrical power. Companies now must operate under a mandate that says the average customer may lose power for no more than one second per year. In order to manage their infrastructure, electrical companies must invest in new technology.

Similarly, other mandates could spur similar investment. Dr. Zimmermann offers the example of biofoods, or organic foods. People are willing to pay premium prices for organic foods. But how can shoppers be assured that these foods truly meet organic standards set by the government? The IOT could provide a pedigree of these foods and document its journey from farm to foodshelf.

"It's the indirect effect," says Dr. Zimmermann. "When you give some more consumer rights on what we all buy every day, you have to have an Internet of Things to implement it and bring a return on investment in a profitable way."

9

The technologies of the Internet of Things open the door for revolutionary gains in efficiency, productivity, and micromanagement. Tools like RFID, GPS, and digital video cameras give managers much more detailed insights about what is happening in their own organizations and a greater ability to control fine details. This new form of management has been termed "high-resolution management" (HRM). Its greatest champion is Prof. Fleisch of ETH Zürich and St. Gallen University.

Prof. Fleisch believes the IOT will transform the very nature of management. It arms managers with more accurate diagnostic tools and allows them to toggle freely from macro to micro. It increases the magnification at which they can measure, plan, and act. As a result, managers can break their operations into smaller fragments and take a finer-grained approach to markets, products, points of distribution, and supply chains. To make an analogy with an athlete, high-resolution management gives the enterprise quicker reflexes. It shortens the reaction time between perception and action. It makes companies faster, more agile, and more competitive.

## Cheaper Technology, Richer Data

According to Prof. Fleisch, today's Internet is an "isolated island" of computers and mobile phones that connects to the real world by manual input. He compares this to the "stone age of computing." As he sees it, today's computers have no eyes and ears and give us only a blurry, low-resolution image of the world.

But emerging technologies give us a better picture, like a digital photograph with more pixels and higher resolution. In the IOT, smart items become nerve endings that can report to anything and anywhere. Every object in the real world can have its own address or a "virtual twin" in the digital world. Eventually, every object can become part of the Internet. At any given moment, managers can take a high-resolution snapshot of their organization

These advances will drive new business innovation. When the costs of real-world sensing are high, companies tend to use it sparingly. They might check inventory once a year or tag only large items, like cargo containers. But when sensing becomes more economical, Prof. Fleisch said, companies want to do it all the time. They want to monitor inventory continuously, expand RFID from large containers to single items, and monitor how items interact with surroundings. Companies start to produce richer and more voluminous data. They begin to view processes—and for processes to read themselves—in real time.

As Peter Drucker famously observed, you can manage only what you can measure. With the IOT, one can measure, and thus manage, infinitely more. Prof. Fleisch calls this "real-world mining." This leads not only to a richer plethora of data, but also to more reliable data. Because this data is quietly and continuously collected over time, it becomes harder to influence or distort. It becomes "trusted" and "honest" data. For example, sensor-collected delivery data is far more reliable than asking truck drivers to fill out a questionnaire about whether they have been on time.

Prof. Fleisch believes that all this data will change the nature of management itself. In the old days, companies had less real-time data and thus had to devote more time to planning. Since they couldn't see a reliable picture of operations, they had to imagine what the picture might look like. This took the form of planning, budget forecasts, contingencies, risk management, and so on. With the IOT, real-time data closes the gap between planning and action. Companies can plan less and manage more.

"There is not a big gap between planning, doing, and checking anymore," he said. "Doing, checking, and planning are growing together. So it's more trial and error."

## The Evolution of HRM: Lessons from the Auto Industry

High-resolution management can be seen as a continuation of business trends that have been underway for more than a century.

According to Prof. Fleisch, the IOT rests on a fundamental economic principle: it dissolves the transaction costs of crossing the "media break" between the real world and the virtual world. In fact, he portrays the history of computing—from the punch card to the keyboard to the barcode—as one continuous saga of abolishing these media breaks. An automated accounting system allowed enterprises to enter data only once. Barcodes saved the trouble of having to manually log an item again and again. ERP systems dissolved many media breaks with company-wide information systems. Now the IOT, with technologies that bridge the last mile between the physical and virtual world, removes another barrier between the physical and digital worlds.

According to Prof. Fleisch, the automotive industry provides a revealing example of this evolution.

The auto industry began as an artisanal trade in the 1890s. Cars were built in artisanal workshops of highly skilled workers who used general machine tools. These companies made no more than 1,000 cars per year and no more than 50 were alike. Cars were customized to the tastes of customers. In 1908, Henry Ford changed all that by introducing mass production of the Model T. The Model T had two primary objectives: repeatabilty of production processes ("Any customer can have a car painted any color that he wants, so long as it is black") and ease of use by anyone (previous cars often required chauffeurs and mechanics). Mass production emphasized a consistent methodology for interchanging and assembling parts. Three years later, Frederick W. Taylor introduced the principles of scientific management, which emphasized scientific study of tasks, methodological selection, training, and supervision of workers. Ford and Taylor used stopwatches and notepads to conduct time and motion studies to improve the efficiency of their processes. These models dominated the US auto industry for decades, as it rose to dominate the global market.

The next major evolution in the auto industry occurred in the post-war years in Japan. Led by Toyota Motor Company, "lean manufacturing" emphasized four essential principles: teamwork, communication, efficient use of resources (including the elimination of waste), and continuous improvement. These approaches rested on the visionary thinking of Toyota executives Eiji Toyoda and Taiichi Ohno and US management theorist and consultant W. Edwards Deming. Once again, these breakthroughs relied on innovations in measurement: time and motion studies, continuous improvement, statistical tracking charts, sampling principles, generating statistically sound conclusions, and the "measure-plan-act" loop. They introduced now-familiar terms such as kaizen (constant process analysis), kanban (pull production based on the needs of the final consumer), and poka yoke (mistake-proofing).

These practices fueled the rise of the Japanese auto industry and allowed it to grow from an industry widely perceived as manufacturing cheap "tin cans" to one that threatened to knock Detroit off its perch. Between 1960 and 1983, Toyota's net added value per employee nearly tripled from $26,039 to $73,897. Meanwhile, Ford's added value per employee barely increased from $31,272 to $37,235. In 2006, Toyota surpassed General Motors as the world's top auto manufacturer. These innovations helped Toyota dethrone Detroit.

They also transformed the art and science of management itself across many industries. Now statistical process control is a staple in modern manufacturing and process engineering. A factory for microprocessor wafers might track hundreds of parameters in real time and make adjustments to ensure high quality and yields.

According to Prof. Fleisch, high-resolution management is a continuation of these trends. The new twist is more powerful tools like RFID, GPS, and digital cameras. In one sense, these tools are revolutionary; in another they are evolutionary in that they advance trends that began a century before with Ford and Taylor. As Prof. Fleisch and two colleagues observed in one paper, "High-resolution management is a natural evolution that is based on prior principles of measurement, root cause identification, real-time control, and run-to-run improvement. Just like its ancestors—craft, mass, and lean production—high-resolution management is bound to have a profound effect on every aspect of an organization."

## Drilling Down: Applying HRM

The IOT and high-resolution management will profoundly change the nature of the enterprise. Managers can easily switch from macro to micro levels. They can look at the big picture one moment and the next moment drill down to the finest levels of granular detail. They can measure, plan, and act with greater knowledge.

Many businesses simply lack a detailed picture of what is happening within their own walls. For example, companies simply lack the ability to track their inventory in detail. One Harvard Business School study of a major US retailer found that 65% of inventory records were inaccurate.

Consider a shipment of razors. At the factory, each razor is put in a package with several others, boxed into cases, grouped onto pallets, loaded into a container, and shipped to a manufacturing distribution center. At the manufacturing distribution center, they may be broken down into pallets, or even cases, and sent to the retailer's distribution center. There they may be further grouped or broken down into containers, pallets, or cases—adding more confusion to the picture—and shipped to the retail stores. At the stores, they are finally broken down into packets again.

This whole system is confusing and vulnerable to error. Manual inventory of scanning barcodes is costly, time-consuming, and cumbersome. Imagine if one four-pack accidentally was coded as an eight-pack: this error would be amplified through the chain and accumulate with other errors. In short, said Prof. Fleisch, the current approach lacks the resolution to spot errors and inefficiencies.

Now imagine the same processes with RFID. RFID uses a passive tag and does not require a person to scan each barcode. Instead, it just reads the tag as the item passes by. In high-resolution management, managers can drill down to minute detail.

Prof. Fleisch argues that high-resolution management increases the magnification at which businesses can operate so a finer level of detail can be understood and orchestrated. It allows greater fragmentation of their markets, products, points of distribution, and supply chains. Companies could greatly reduce problems like theft, shortages, or overstock.

These approaches already have allowed some industries to refine their business models into smaller-grained fragments. For example a news web site contains dozens or hundreds of services—stock quotes, weather, online advertising, predictive modeling, a number of online retailers, and a digital notary. Such fragmentation will become increasingly common even for simple transactions and businesses.

The IOT does not necessarily require a wholesale, system-wide adoption. Sometimes companies can reap the benefits by applying the IOT technologies in discrete situations and then applying the lessons more broadly. For example, Prof. Fleisch recounted how the retail store Metro tagged pants and found that jeans on hangers sold better than those on shelves. Until then, people debated which form of presentation was more effective, but nobody really knew. The store gained valuable insights that could be applied elsewhere.

"This is really kind of a tool to learn the reality of your business processes and improve the business processes," he said. "It's a consulting tool."

Others have made similar observations. Dr. Christian Floerkemeier, and his colleagues at the Auto ID Labs at MIT recently examined the execution of a product promotion. Ideally, a product launch is accompanied by an advertising campaign, promotional displays in stores, and a release date when retailers are supposed to place the new product on the shop floor.

In this case, the displays were equipped with RFID tags. When they looked at the data they found striking proof that the execution left much to be desired.

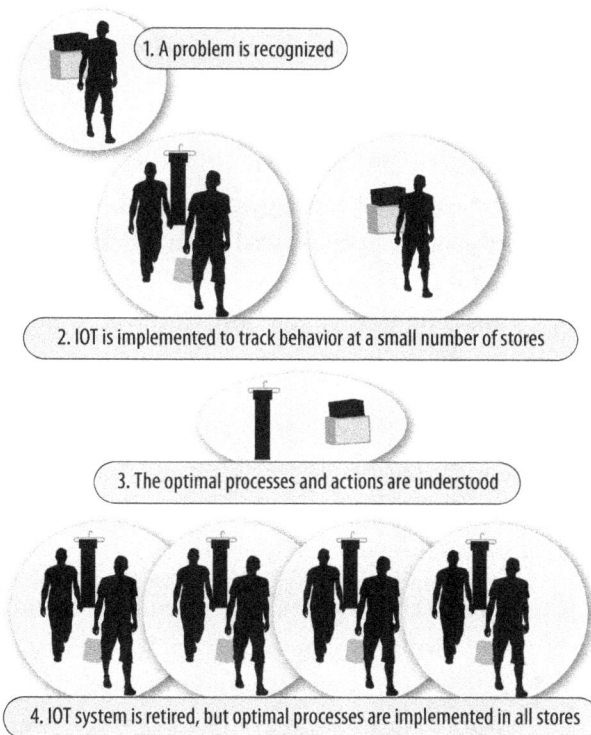

1. A problem is recognized

2. IOT is implemented to track behavior at a small number of stores

3. The optimal processes and actions are understood

4. IOT system is retired, but optimal processes are implemented in all stores

*Figure 9-1. Using the IOT as a Diagnostic Tool*

"Hardly any of the shipments arrived in time for the promotion launch date in television commercials," Dr. Floerkemeier recalled. "Many of them came after the date. Many of them actually arrived in the stores a couple of days earlier."

Normally this sloppy execution might have gone unnoticed. But in this case, the instrumentation brought the problem to light. As Dr. Floerkemeier said, "Better mapping between the real and virtual world allowed you to see that something was going fundamentally wrong here—some insight that the brand owner wouldn't have had before."

Better yet, they could actually diagnose where the processes went wrong. As soon as stores received the promotions, they put them on the shop floor. The instructions to hold them until a certain date proved too complicated. High-resolution management enabled them to identify potential solutions. They could see that the items were placed on sales floors four to eight hours after they arrived. If shipments arrived on a certain date, they were almost guaranteed to go on display that day. As a result, managers learned they could control the display dates by altering their shipping schedule.

This example captures the essence of high-resolution management and the IOT. "You can now actually see where something works or doesn't work, and you can adjust your processes in a much more flexible, much more decisive way," said Dr. Floerkemeier. "Previously it probably would have taken you a long time to find that out. You would have to visit all the stores and find out how displays are being put on the shop floor."

## The Lessons

How to begin reaping the benefits of high-resolution management? Prof. Fleisch has identified several principles. The following best practices are drawn from the experience of early adopters:

- **Try Complex Problems:** Nuts do not require a sledgehammer. You do not use MRI to diagnose a scratch. The best applications of the IOT deal with complex problems, such as management of large supply chains and are hard to implement. The IOT is not a good option for simple problems that can be worked out through simpler means

- **Search for Blind Spots:** Look for things you didn't know were there. Companies often use IOT data to make unexpected discoveries. They set up IOT technologies to address one problem and generate data that illuminates something else. This is analogous to a team doing an X-ray for a broken bone and unexpectedly discovering a tumor. High-resolution data helps enterprises identify things they couldn't see before

- **Diagnose and Improve:** It is difficult to build a business case for IOT because it is impossible to calculate the return of an investment when the problems to be solved are partly unknown. As a result, some companies introduce IOT technology as a simple diagnostic tool and later use the extensive data to improve their processes

- **Automate Low-Level Management:** Many companies use IOT technology to automate simple manual tasks, such as signing in cargo, updating inventory records, or sending notifications. They eliminate tedious tasks that previously required humans and save time and labor

- **Measure, Manage, and Innovate:** At the highest level of innovation, companies use IOT to transform their business models. For example, they might change their business model from selling a product to renting it on a per-use basis

## Barriers

Like much of the IOT, high-resolution management must clear a number of hurdles before it becomes common practice. Many of the challenges of high-resolution management are the same as those faced by the IOT as a whole, outlined earlier in this book.

### The Unknown Business Case

One barrier is the unknown: how can you measure the benefits of finding something you can't yet see? This is like trying to justify a diagnostic procedure like an X-ray. Companies need to take it as an article of faith that these systems will bring insights that ultimately justify their cost. Early adopters tend to be companies with a strong CEO who believes in the technology.

"You search for blind spots," said Prof. Fleisch. "You can't make a business case using a problem you don't know. You can't make a business case when you search for blind spots."

### Filtering Data

Another challenge is sifting through information. What is vital and what is just white noise?

### Scale

Scale is a major challenge. For a large company like Wal-Mart, attempting high-resolution management is a huge undertaking because it encompasses so many business processes, inventories, and partners.

Recently, companies have been deploying it more surgically, to discrete business processes and products, rather than attempting to do it across their whole supply chains. People don't necessarily need to suddenly instrument their whole supply chains. Instead, they can take a more pragmatic and limited approach.

"It's difficult to do this on a very grand scale," Dr. Floerkemeier says. "People are much more careful in picking certain sectors and certain problems that they're trying to analyze. They're moving away from trying the IOT in open-loop supply chains and moving to very specific approaches, limited in scale."

**10**

The Internet of Things is not just a matter of hardware. To see it as just a collection of devices is short-sighted. The more important question is this: what is enabled by these devices? How can we use these tools to become smarter, faster and more agile? How can we leverage the IOT into new business processes and new models? In this chapter, we examine how the IOT is being used in real-world applications and spawning new business models.

## Emerging Models

Smart objects have the potential to overhaul business processes and create new business models. The devices themselves are ready. But one major barrier is the lack of proven business models.

"To make it work, we need collaboration between several companies," said Prof. Österle of the University of St. Gallen. "We have to get sensible business models for not just one, but five or ten companies working together. That makes things much more complicated."

How do we set up processes among many different partners? How do we define and divide revenue streams? How do we condense the signals into meaningful data? Prof. Österle says many layers need to be defined: technical, data extraction processes, and business models. Even now, the technologies are still maturing.

According to Prof. Österle, the absence of business models is a larger problem in the business-to-consumer realm. In the B2B realm, there is more precedent for collaborative processes and clearer definition of revenue streams. "Consumers expect all these services to be free," he said. "This is totally different from B2B."

His work has highlighted a number of potential barriers. One is the basic investment. Another is that the return on investment may seem small because revenues are divided among so many partners. Often there is a long period with little revenue, just as we saw in the early days of Internet startups, many of which went under. Look at how long it took Amazon and eBay to become profitable.

Much of the discussion of the IOT focuses on technical aspects. But Prof. Österle believes the larger problem is a lack of understanding about how to create consumer processes and business models. How should we share the costs of development? How should we divide revenues? Who should orchestrate these systems? And, how can you achieve the market size to make them viable?

To make the business model work, we often need the collaboration of multiple companies working together. But Prof. Österle takes heart from precedents like the Internet travel industry. In the 1990s, Internet applications started to revolutionize the travel business by integrating hundreds of thousands of players: airlines, hotels, car rental agencies—all accessible from a single web site.

In the beginning, most of these Internet sites started as small, narrow applications. Even the mighty eBay started as a platform for exchanging Pez containers. Sites grew and some were so successful that they turned into colossi like eBay or Amazon.

"We have to come to a similar solution in the Internet of Things," Prof. Österle said. "It's not a matter of the Internet of Things, it's a matter of consumer processes. I think the evolution will not be so different from what we saw with web sites like Expedia."

## New Data Builds New Models

Indeed, we are seeing examples of new business models emerging. Many of these remain nascent and confined to localized applications and closed loops. Yet this is a natural progression for emerging technologies and we are likely to see these models mature and give rise to new ones.

The devices of the IOT bring more granularity of data. Prof. Fleisch of ETH Zürich and University of St. Gallen divides pervasive computing into two categories. The first class changes how companies manage their resources but does not alter their basic business models. Most existing RFID projects fall into this first category of incremental process improvements—applications that seek to reduce failure rates, shrinkage, operating stock, on-time delivery, and so on.

The second class brings more radical change. Companies use this new stream of granular data to unleash far-reaching changes and new business models. Prof. Fleisch points out that high-resolution management—made possible by the devices of the IOT—shows companies what drives or destroys value. Using this information, they can transform their business model, say from selling services at a fixed price to usage-based pricing. For example, a car insurance company might collect data on customer driving patterns and recalculate car insurance premiums each month based on factors such as day and night driving, weather, highway, or neighborhood. According to Prof. Fleisch, this second approach allows companies to "strive for more radical innovations that affect customer relations, services, and revenues models."[3]

In the following sections, we will examine how technologies like RFID are giving rise to new business processes and new business models in areas like digital product memory, logistics, manufacturing, services, anti-counterfeiting and smart energy.

---

[3]  Fleisch, Elgar, "Business Impact of Pervasive Technologies: Opportunities and Risks," *http://www.autoidlabs.org/uploads/media/AUTOIDLABS-WP-BIZAPP-018.pdf*

## Digital Product Memory

Someday every product may carry its own black box.

In the IOT, products have the ability to carry information about themselves. With RFID, these products can record their own life history, known as digital product memory (also called semantic product memory). They can tell us the raw materials that went into making them, where and when they were made and transported, and what items they encountered. They can tell us about the environmental conditions, such as time, temperature, or humidity. In short, they are equipped with a digital memory.

Prof. Wahlster of DFKI and Saarland University calls semantic product memory a killer application of the IOT. "There are many, many different applications, and every day more and more people are interested in this area," he says.

Prof. Wahlster offers many examples that are tested in the SemProM project, a large consortium funded by the German Ministry of Education and Research (BMBF) with more than 16 million euro.

Personalized weekly blister packs of medicines might help a person keep track of dosages, remind them to take their medicines four times per day, and make sure they are not combined with any other drugs that would produce toxic combinations (see: *http://www.7x4pharma. de/international/index.php*).

A "black box" inside a car could allow a person shopping for a used car to use her Near Field Communication-enabled smart phone to check its maintenance records and discover it has inferior third-party brake pads. Similarly, the digital product memory could help enforce warranty repairs—or void service guarantees if the owner neglected to change the oil at prescribed intervals.

Products may even talk to each other. Cars may form a network of roaming nodes. One car may warn another car traveling in the opposite direction about icy road conditions one kilometer ahead with Car2X technologies tested already in the SIM-TD test field around Frankfurt (see: *http://www.simtd.de/index.dhtml/344b3f460b571450063h/-/ enEN/-/CS/-/*).

Even items in the grocery aisle might keep a diary. Today, a frozen pizza box might contain simple information like a barcode, a list of ingredients, and perhaps an organic certification. In the future, an RFID tag might contain a plethora of digital information. It might carry a detailed breakdown of every single ingredient in the product and its entire life history. It could show where and when the tomatoes and wheat were grown, where and when the cheese was made, and where and when they were processed, transported, and stored. It might confirm that the frozen pizza was stored within a specific range of temperatures. All of this information would be available to the consumer in the grocery aisle via a mobile phone.

"There are so many benefits of product memory that we believe it's really one of the killer applications," said Prof. Wahlster.

Applications can grow more complex. Airliners have long carried "black boxes" that record the flight data and voice communications of each flight. In the past, the high cost of these technologies has limited this kind of product memory to high-value items like planes. But advances in technology increasingly allow us to deploy product memory to items of lesser and lesser value. Someday, they may become as commonplace as barcodes are today.

This opens the door to many new applications—supply chain management, anti-counterfeiting, customer relations, and new services tailored to individual customers. We will discuss some of these examples elsewhere in this book.

According to Dr. Rode of SAP Research, early examples of digital product memory have been emerging for years. In factories, barcode, data matrix code, and RFID are often used to track movements and status of containers, equipment, and tools, or to carry data for process control, process safety, quality management, or material replenishment (eKanban). "It hasn't been called digital product memory, but that's exactly what it is," he said. "These are small and focused applications and that's why it works. Today, it's mainly based on barcodes but it's being replaced more and more by RFID."

Hospitals might use technologies to keep track of medicines or even patients. "These domains are very closed loops and well-understood problems," said Dr. Rode. "The reason it hasn't been taken up in the open loops is that the infrastructure isn't there."

Until now, the deployment has been curtailed because of lack of standards and a lookup infrastructure. But if these can be resolved, he sees digital product memory as a potential open-loop application of the IOT.

Dr. Ackermann of SAP Research, agrees. He predicts that digital product memory will take root in closed domains like the manufacturing shop floor. Over time, it will expand to the entire logistics chain. For example, an auto company could keep track of parts through the whole product lifecycle from production to customer use, and even up to dismantling and recycling.

"These higher-level applications that operate on larger groups of products, operate on longer time spans, operate on scenarios where even ownership changed over time, they are enabled by the semantic product memory," Dr. Ackermann said. "This is a building block. You have to build on top of it, but you benefit from what is done below."

He foresees a layering of information-carrying systems. It will start with passive tags and progress to active tags, small embedded systems, and other advanced technologies.

Product memory allows companies to develop new business models. The histories of product classes can be aggregated and analyzed. What are usage patterns? What are failure rates? How do high-level business processes fit with low-level usage? By looking at these patterns, companies will be able to refine their business processes and design new ones. For example, an equipment provider may gain new insights about how customers use the machines and how often they have to make repairs or replace parts. Based on these insights, the company might convert its maintenance contracts from a flat rate to a pay-per-incident model.

These insights can be further enhanced by semantic technologies that bring together information from many sources. For example, they may learn about product failure rates from user forums or blogs, and feed this information back to the designers and manufacturers so they can fix the problems. These "semantically enhanced information federation systems" can establish feedback loops that allow companies to correct and improve processes with real-time information. (These semantic technologies are the focus of the Aletheia project discussed below.)

Dr. Ackermann predicts that these applications will expand over time. Once there is critical mass of adoption, another business case will be added.

"One of the benefits here is that there might be business cases that pay off really fast, even limited ones," he says. "They don't necessarily

## A Brief Primer on DPM Technologies

Product tags fall into three basic categories:

- **Class One:** No information processing on the product, only identification and reference to external storage. Examples are barcodes and RFID

- **Class Two:** The products are capable of storing sensor data and communicating with other tags. New technologies of this class include Sun's Small Programmable Object Technology (Sun SPOT) and Intel iMotes

- **Class Three:** These advanced tags are capable of semantic integration of data and semantic query processing. These are called "Next Generation Smart Items." There are no commercial products of this type yet

(See *Figure 5-4* on page 61)

SemProM has developed a prototype standard for all three classes of product memories. According to Prof. Wahlster, the SemProM working group hopes to have their standard ratified by the World Wide Web Consortium.

"EPCglobal is subsumed by our standard, because it's very weak," said Prof. Wahlster. "They have no notion of checking the environment of the product and the context. There is no sensor information involved. Everything which is in EPCglobal is in our semantic product memory standard, but it's much more."

have to be global ones with billions of items. Just limited ones can really pay off and make companies invest in this technology and spread it."

Prof. Wahlster is optimistic about digital product memories. Initially, many companies rejected RFID because of the added expense and because they saw little advantage over existing barcodes. But he believes they will adapt product memories once they see the added value and the opportunity to extend new service offerings.

According to Prof. Wahlster, 90% of processors are now going into embedded systems. In other words, they are not going into traditional computers like laptops or PCs, but into everyday items like washing machines or cars (indeed, a single BMW has 70 embedded computers!).

"Now we really have the case that we have so much added value that we think this will fly, and more and more people are jumping on it," said Prof. Wahlster. "I'm afraid it will be another five years minimum to have a complete breakthrough. We will see what happens, but it looks very promising."

## Research Initiatives

Several research initiatives are underway to bring this vision closer to reality. The German government has invested 45 million euros in the IOT. Three research projects in particular explore digital product memory: SemProM, Aletheia, and ADiWa. SAP Research is playing a major role in each of these initiatives.

### SemProM (Semantic Product Memory)

SemProM focuses on the data-level connection of objects in the real world. SemProM explores how to use product data from devices like 2D

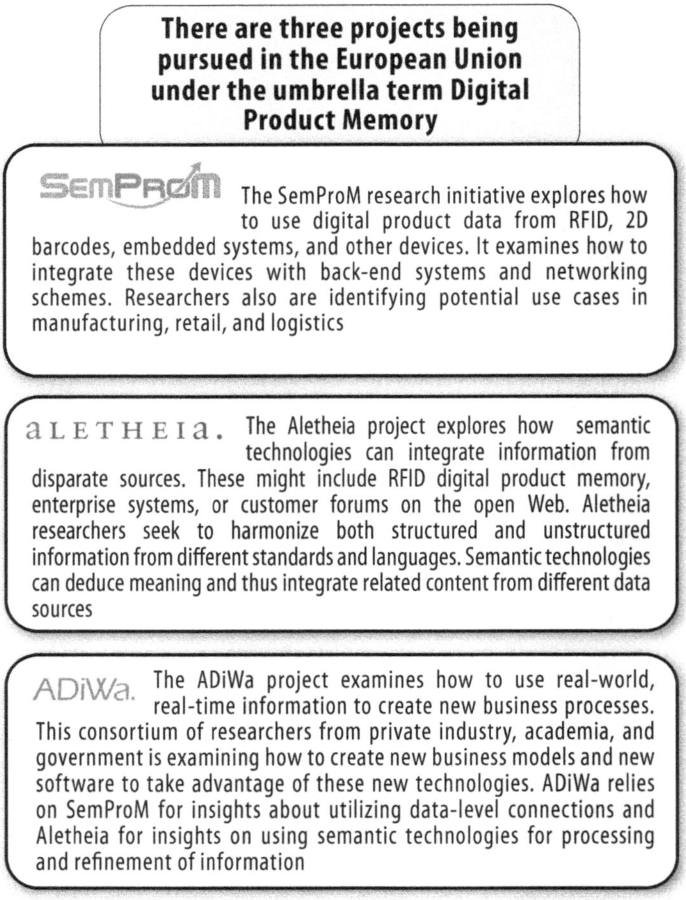

**There are three projects being pursued in the European Union under the umbrella term Digital Product Memory**

SemPro M The SemProM research initiative explores how to use digital product data from RFID, 2D barcodes, embedded systems, and other devices. It examines how to integrate these devices with back-end systems and networking schemes. Researchers also are identifying potential use cases in manufacturing, retail, and logistics

aLETHEIa. The Aletheia project explores how semantic technologies can integrate information from disparate sources. These might include RFID digital product memory, enterprise systems, or customer forums on the open Web. Aletheia researchers seek to harmonize both structured and unstructured information from different standards and languages. Semantic technologies can deduce meaning and thus integrate related content from different data sources

ADiWa. The ADiWa project examines how to use real-world, real-time information to create new business processes. This consortium of researchers from private industry, academia, and government is examining how to create new business models and new software to take advantage of these new technologies. ADiWa relies on SemProM for insights about utilizing data-level connections and Aletheia for insights on using semantic technologies for processing and refinement of information

*Figure 10-1. Research Projects Related to Digital Product Memory*

barcode, RFID, and embedded devices in domains such as manufacturing, logistics, or retail. This effort integrates technologies such as existing system hardware, networking systems such as RFID, semantics, machine-to-machine communication, intelligent sensor networks, and multimodal interaction.

The SemProM project already has identified more than 50 potential use cases. A few examples:

- Manufacturing
  - ▫ Decentralized production control: routing items based on stored information
  - ▫ Localization: tracking and tracing of production orders, materials, tools, machines, and personnel
  - ▫ Robust production: in-process quality checks and ability to intervene if pre-defined rules are violated
  - ▫ Maintenance and repair: customer service and warranty support
- Retail
  - ▫ Item integrity: using digital product memory to comply with external and internal guidelines and provide product information
  - ▫ Product recommendations: in-store product suggestions communicated directly to customer's mobile phone

### Aletheia: Improving Access to Product Information through Semantic Technologies

The Aletheia project (named after the Greek goddess of truth) uses semantic technologies to harmonize product related information from a variety of often-disparate information sources. These may include back-end enterprise systems, emails, or Web 2.0 resources like blogs, wikis, or public forums. The challenge is how to link related content that may reside in different systems and use different languages and standards.

As it now stands, the Web lacks this kind of semantic intelligence. If you search for the keyword "car" you will get information that contains that keyword but not necessarily files that use the term "automobile."

The system does not recognize that these two terms refer to the same thing. It does not recognize implicit meaning.

In the future, enterprises must integrate data from an array of sources. This includes structured data from ERP systems, unstructured data from office documents, IOT data like RFID data, and even public data from the open Web such as wikis, blogs, and web forums.

Due to the heterogeneity of these sources, this information must be converted in order to be useful to a query. Semantic technologies can help deduce meaning and thus make relevant information available to the user.

### ADiWa: From the IOT to Intelligent Business Processes

The ADiWa ("Allianz Digitaler Warenfluss," Alliance Digital Product Flow) project explores new ways to access information stored in the digital product memory. ADiWA is a consortium of researchers and companies from business and technology who are examining how to use real-world, real-time information to create dynamic new business processes. It opens the door to new SaaS business models and represents a symbiosis between the IOT and the Internet of Services.

The full potential of the IOT can only be realized with real-time, automated processing of information and business-related events. Products pass through many different stages, such as production, transit, sales, maintenance, and so on. This requires us to create bridges from myriad information systems and recognize complex events. These processes need a new, complex software logistic system.

The ADiWa project investigates potential mechanisms to solve this problem. It draws from both SemProM (for insights about utilizing the data-level connections) and Aletheia (for insights on using semantic technologies for processing and semantic refinement of this information).

## Future Applications

Digital product memory is not limited to simply storing information. It also can transform production processes. At the DKFI's Robotic Innovation Centre (RIC) headed by Prof. Frank Kirchner, researchers are developing a robot capable of reading the digital product memory of each item it encounters. This project is part of the SemProM research initiative.

Logistics processes must adjust their procedures for each item they encounter. A durable and heavy item, like a pallet of bricks, requires a lot of brute force but not much delicacy. A shipment of glass requires not much strength but a lot of finesse. How to distinguish between products with very different handling requirements? Digital product memory can store handling information in machine-readable form.

Enter Mr. SemProM. The robot has a flexible head and two arms, each with seven degrees of freedom and gripper hands. Its head has stereo and 3D-cameras for object recognition and navigation, and two computers for image processing and manipulation tasks. It also has another camera in its gripper for fine motor movements and an RFID reader to scan each item it encounters.

The robot will read each product memory and treat the item accordingly. The robot's grippers automatically adapt to specifications such as size, weight, or proper lifting points. It also guarantees that constraints like maximum acceleration or geometric orientation are not violated.

These approaches are particularly valuable in production lines that produce non-uniform items. In these environments, the constant variation of products and processes puts a premium on quality control, flexibility, and instant adaptability. Traditional control methods are insufficient.

We can also imagine how digital product memory may transform more complex processes. In these cases, multiple products, tools, backend systems, and people may all talk to each other.

Consider the example of aircraft maintenance. The maintenance, repair, and overhaul (MRO) of aircrafts are essential requirements for safety and compliance with government regulations. These processes are heavily regulated, standardized within the industry, and require detailed recordkeeping. Passenger airlines must perform base checks approximately every 650 flying hours and MRO accounts for about 12% of the total operating costs of an aircraft.

As it now stands, existing methods are undermined by inefficiency, inadequate tool management, human errors, and labor-intensive manual documentation and check procedures.

In one scenario envisioned by Prof. Fleisch and his colleagues, a ubiquitous computing environment could create a new model for aircraft maintenance. This scenario is described in the paper "A Ubiquitous Computing Environment for Aircraft Maintenance" by Matthias Lampe, Martin Strassner, and Elgar Fleisch. Every mechanic would carry a personal Pervasive Device (PD) that acts as an interface to the system architecture and contains all the applications the mechanic needs.

At the beginning of a MRO task, the PD gives the mechanic a work order with details about the assignment. The system automatically generates a list of tools and parts. The mechanic does not have to manually request these items; instead, the mechanic just picks them up. The PD reads the RFID signature of each part and tool and verifies they are the right ones.

During the job, the PD guides the mechanic through the steps. It displays the relevant sections of maintenance handbooks and checklists. The mechanic must confirm each step manually or by voice command.

For each part, the PD displays the digital product memory with the maintenance history and status. The mechanic updates the status of the parts by dictation to the PD. This information is automatically logged in the official report.

At the end of the job, the system creates an electronic certificate of release-to-service-and-maintenance statement with the digital signature of the inspector. The PD prompts the mechanic to return the tools to his toolbox. If a mechanic mistakenly returns a tool to the toolbox of a coworker, he immediately gets an alert on his PD. The device also directs mechanics to return tools to the inventory and confirms when they are dropped in the return box.

This scenario offers many potential benefits: reduction of delays and errors, automated documentation, and ease of use. Of course, there also are challenges to realizing this vision. RFID tags suffer from radio interference when attached to metal objects or in industrial areas with lots of machines and electromagnetic activity. Special steps must be taken to create more robust tags that work in these environments. Ubiquitous computing lacks standards and needs further integration. Such a scenario is feasible if all part and tool manufacturers use the same standards for product identification.

## Supply Chain Management

Supply chain management represents one of the most promising applications of the IOT. This was made apparent at the IRF 2009 conference when Prof. Min of Fudan University touted supply chain management as a killer application of the IOT. This potential was borne out by the further research conducted by the authors of this book. In the discussion of high-resolution management, we saw how the IOT could enhance efficiency and reliability and accuracy of supply chain logistics.

## The Vision: The Transparent Supply Chain

RFID has many advantages over barcodes. RFID does not require line-of-sight and tags can be read from a distance. Many tags can be read at once, improving speed and efficiency. No manual scanning is required, reducing labor costs. It improves visibility and traceability. It potentially allows tracking of individual items, which could revolutionize supply chain management.

The following are some major benefits.

### Labor

RFID can help automate the supply chain. At a typical distribution center, labor accounts for an estimated 50 to 80% of costs. One study predicted that RFID could reduce check-in time by 60 to 93%, order-picking labor by 36%, and verification by 90%.

Or consider how it might speed transportation logistics. Truck drivers delivering a load of grapes to a winery no longer would have to park, get out of the cabs and fill out paperwork, and get tickets for each delivery. Instead, drivers could simply pull onto a scale while an RFID reader instantly pulled down all essential details from the tag and continue on their business.

### Visibility

RFID provides telescopic and microscopic visibility into supply chain operations and always remains on. Consider the monumental logistical challenges of tracking thousands of shipping containers across the world. RFID can automatically track the movements of every single container, a history of all its shipments, and even a record of what containers moved alongside it.

It also can highlight inefficiencies. One study estimated that the US retail industry loses about $70 billion per year due to inefficient supply chain practices, almost half due to products not being on the shelf when consumers look for them. The real-time visibility of RFID could drastically reduce this problem. Indeed, some have envisioned "smart shelves" with built-in RFID readers to keep constant tabs on inventories.

### Item-Level Tracking

At this point, most RFID users tag mostly large items like pallets or shipping containers. In the future, however, the ability to tag and follow single items opens up a world of new possibilities—constant inventory management, theft detection, recall management, or warranty records. Someday, RFID tags may even enable automatic checkouts.

### Traceable Warranties and Recalls

Similarly, recalls and warranties may be managed with greater accuracy. Say a car manufacturer produces a model that uses a part from several suppliers. When one brand proves defective, can the manufacturer reconstruct which cars have the bad tires and which ones don't? Companies will no longer have to take a scattershot approach in which they recall every product that might be affected. Instead, recalls will be rifle shots where individual items can be targeted with precision.

### New Insights

Instrumenting the supply chain is not only a matter of tracking shipments. It also can yield new forms of data and new insights.

In some cases, valuable information comes from interactions of the devices themselves. A few years ago, Ms. Murphy-Hoye and her colleagues at Intel instrumented a cargo ship in the Pacific in a pilot study. They put sensors both inside and outside the containers and captured data on temperature, humidity, and light, as they sat in the ship's hold. They installed RFID readers and light sensors inside of the cargo hold doors so that when cargo entered or exited it would read RFID tags of the things that went in and out of the cargo space.

The Department of Homeland Security wanted to figure out how to detect a 9-square-inch hole in a cargo container, but couldn't figure out a way to do it without an ungodly number of sensors.

But Ms. Murphy-Hoye's research suggested there may be another way to protect each container: the herd effect. The sensors create a network and record of contacts, all of which can yield further insights. "They can talk to each other," said Ms. Murphy-Hoye. "It's a mesh, it's self-configuring. If I have the cargo containers talk to each other, I'm creating safety in numbers."

These mesh networks can reveal whether one container might have been removed from the group or separated sometime during the loading process. "You can even reassemble the route of an object based on what other objects heard about it along the way," says Ms. Murphy-Hoye. "It's like the invisible man, right? You know that invisible man is brushing his teeth because there's a toothbrush in the air moving up and down or you know he has left a room because the door opens and closes. It's activity inference based on the things around you and the responses of those things around you."

These mesh networks also suggested new methods for creating a secure envelope. According to Ms. Murphy-Hoye, these networks created a radio signature inside the cargo container. Any disturbance—even opening the door to the cargo hold—caused a discernible change in this signature. She speculates that further research could help us better identify what types of actions would cause certain changes in signatures. "You can create a dynamic signature, fill your cargo container with radio signals, and solve the problem. Radio could create the envelope of safety and security for the cargo container."

Interpreting this data will be a major undertaking in the future. "This is all about proactive computing and proactive enterprise," says Ms. Murphy-Hoye. "You're turning around the way that you think about your ability to not just respond but to determine the best course of action."

## Barriers: Security and Trust

There are many barriers to realizing this vision of a transparent supply chain. How can we protect these devices from attack? How can these networks be expanded to include more players without compromising security and privacy of data? How can participants be confident that other companies in the logistics chain will not snoop on them?

"What is new in the Internet of Things, it's not one administrative domain and it's not monolithic," said Mr. Neidecker-Lutz of SAP Research. "Most of the interesting things happen if you have to do complex event processing across boundaries and you don't want to share the data. There it gets really, really nasty."

The data sits in different systems. Bringing them together is technically challenging but feasible. A more problematic challenge is doing this in a way that preserves security. Or as Mr. Neidecker-Lutz puts it: "Can I do the queries I want, get those things that I want, without exposing too much of the innards?"

In a logistics chain, participants will not be comfortable with a Google-like model where all data goes into a single place and can be accessed by anybody. The key is how to create an entity that everybody trusts—or even create a system with no need for such an entity.

"We already know how to not need anybody to trust if data rates, et cetera, are not too high—that's privacy preserving computation," says Mr. Neidecker-Lutz. "But there's a huge challenge of doing this at large scale with high data rates and high data volumes. It remains to be seen whether that can be done."

Until now, many of these applications have been in closed environments where it is easier to exert control. As IOT applications spread to more open loops, this becomes more challenging. How to ensure security and trust?

Dr. Lotz researches security and trust at SAP Research. He and his colleagues are researching ways to make the supply chain more secure in the IOT.

A multistage supply line requires that only legitimate participants can exchange information. As the IOT moves closer to reality, these chains are increasingly exposed to the physical world and face new vulnerabilities. For example, a hacker could sneak up to a container in a shipping yard or truck lot and use a scanner to read information or try to gain access to the system.

These systems need authentication and integrity of information. According to Dr. Lotz, SAP is investigating some cryptographic computation schemes that would allow authentication and authorized access only to legitimate members. At the same time, they would keep certain data secret, like group membership. Researchers have developed some advanced cryptographic schemes such as a "secret handshake."

"It is not only about authenticating a legitimate participant," he explains. "It is also about not revealing too much information if you are not a legitimate participant."

Privacy friendly schemes are not new; we already have things like identity management and anonymization. But the IOT poses some new challenges in this regard. IOT applications are usually designed for environments and architectures with lots of computing power. The devices have restricted power supplies at the sensors and nodes. Therefore, it is usually not practical to rely on the nerve endings for much of the security.

"It needs to be considered on the application level," says Dr. Lotz. "Typically you are likely to not be able to establish the full set of your security requirements on the level of the things themselves, so you have to take proper precautions on the application level or on the level of your communication and collaboration scenario."

In collaborative supply chains, partners may have qualms about disclosing certain information. For example, a franchise chain may want information to optimize its supply chain, but individual franchise owners may be reluctant to divulge information about their stocks, revenues, or sales. Once again, cryptographic techniques offer solutions. These techniques would allow the system to perform these calculations without disclosing sensitive data to people who are not authorized to see it. According to Dr. Lotz, these schemes use homomorphic properties, randomize inputs, and employ algorithms so that parties do not have to disclose their values to others.

In the next sections, we will examine a few examples of how secure computation may help make these networks more secure.

### Secure Supply Chain Master Planning

Supply chain master planning (SCMP) seeks to optimize operations across the entire supply chain: production, warehousing, and transportation. Full optimization is difficult in a system with multiple partners and limited information sharing.

In theory, it should be possible to optimize a master plan for the entire supply chain, but in practice this ideal is undermined by the guarded, localized approach of each partner. Companies are unwilling to share their private data. They fear that exposing this information might put them at a disadvantage with their partners or that the data may fall into the hands of competitors. As a result, information is hoarded and optimization usually is localized. For example, the manufacturers take one approach that works best for their phase in the supply chain, the transporter and retailers do the same for theirs, and so on. The sum of these localized optimizations does not necessarily add up to systemic optimization.

These obstacles can be removed if each partner has a way of participating with the assurance that their data could not be misused. One potential solution is a shared platform run by a fourth-party, central logistics provider. SAP Research is exploring how secure (Multi-Party) computation (SMC) can be employed such that the relevant data does not need to be disclosed, even to the central planning unit. Each partner would be able to privately and securely submit information with assurance that it would not undermine their competitive position.

In one recent paper, SAP researchers proposed a framework for a secure, centralized supply chain master planning (SSCMP). This paper, "Optimizations for Risk-Aware Secure Supply Chain Master Planning" was authored by SAP researchers Axel Schropfer, Florian Kerschbaum, Christoph Schutz, and Richard Pibernik of the Supply Chain Management Institute. They propose a model that would allow exchange of the data necessary to optimally coordinate manufacturing and transportation decisions, yet still protect the privacy of each member.

Traditional supply chain master planning uses linear programming to centrally compute an optimal production and transportation plan across all parties. In contrast, secure supply chain master planning (SSCMP) uses secure computation and thus protects the confidentiality of input values. As a result, it alleviates the perceived risk for all parties who may be wary of disclosing vital data.

Yet, there is one major drawback of secure computing: higher protection levels can slow down the process. In theory, we could apply the highest protection level of secure computing to all data, but in practice that would make the solutions too slow to be useful. To solve this problem, the SAP researchers proposed a "risk-aware" approach in which input and output data is assessed for risk. What is the potential disadvantage if the data is disclosed? What is the probability that a partner misuses the data to undermine the owner? How much of the data is already generally known? Each level of data can be assigned the appropriate level of secure computation (information-theoretic, cryptographic, or best-effort), cryptographic tools (Homomorphic Encryption, SHA-1, for example), and tool parameters. Not all data is of equal value. For example, disclosing variable shipping costs poses little risk, but data about production quality may be a huge vulnerability. In this model, the most intensive security is reserved only for the most sensitive data.

### Concerns about Industrial Espionage

The use of RFID in supply chains raises challenges about privacy of partners within the supply chain. In some cases, these partners also may be competitors. Imagine two milk companies whose products appear side-by-side on the shelves of the grocery store. Both of these companies participate in the network of the retail store, yet, both are ardent competitors who would love to put their rival out of business. How can these two competitors participate in the same supply chain without risk of disclosing sensitive information to their rivals?

Or imagine a situation in which two companies in a supply chain possess the same tagged item. One company might use the discovery service to get inside information about the supply chain of a rival company. An impostor might seek information about tags—even those he has never possessed—to reconstruct the supply chain of a competitor. Similarly, a counterfeiter might supply false information to mask the origin of bogus products.

To protect against such dangers, we need more secure methods of implementing RFID. An RFID reading generates a tuple (organization, identifier, timestamp), often enriched with additional information, such as reader identifier or type of event (for example, receiving, shipping, unpacking, and so forth). In order to share data, companies must connect with a network such as EPCglobal. This network offers a discovery service that shows a list of all companies that have made contact with a specific tag. One can get a list of all companies to contact about a specific tag and contact them individually for event data. This raises fears of industrial espionage.

SAP researchers Florian Kerschbaum and Alessandro Sorniotti proposed a novel scheme to solve this problem. This approach is described in the 2009 paper "RFID-Based Supply Chain Partner Authentication and Key Agreement."[4] In this scheme, information stored on the tag is tied to a secret key. Only the holder of this private key or "trapdoor information" can prove possession of the tag. As the item changes hands, the tag information is updated with a new re-encryption key or by a trusted third party. The trusted third party also enables the system to trace the item through the supply chain.

This trusted third party can detect illegitimate requests for information. Impersonation is possible only if one party's private "trapdoor" is exposed. A supply chain partner that wants to bring in another party must be willing to share the secret key, relinquish all their business data, and be traceable.

This approach discourages unauthorized disclosure of authentication information by tying it to a secret key or identity. They offered two possible protocols: one based on keys (which allows the trusted third party to hand out re-encryption keys) and one based on identity (which reduces the number of interactions). The authors contend that this solution can work with the simplest current RFID tags in the simplest EPCglobal class and is suitable for wide application. The only requirement is that they must be capable of storing enough information for cryptographic operations.

As the authors conclude, "No rogue user can be successful in a malicious authentication, because it would either be traceable or it would imply the loss of a secret key, which provides a strong incentive to keep the tag authentication information secret and protects the integrity of the supply chain."

[4]  Kerschbaum, F. and Sorniotti, A., 2009. "RFID-Based Supply Chain Partner Authentication and Key Agreement," in *Proceedings of the Second ACM Conference on Wireless Network Security* (Zürich, Switzerland, March 16–19, 2009), WiSec '09. ACM, New York, NY, 41–50, DOI= *http://doi.acm.org/10.1145/1514274.1514281*

## Industrial Privacy and Recalls

We already have seen how RFID can overhaul the process of product recalls. As it now stands, recalls oblige companies to combine multiple sources of information from various ERP systems. They also lack detailed traceability and force companies to err on the side of safety and recall more products than really necessary. RFID can revolutionize this process by storing batch numbers and all parts or ingredients used in all manufacturing steps. It is a more accurate life history of the product.

We hear about these recalls on a regular basis. A food company might be forced to recall items that have become contaminated or spoiled. Drug companies may have to recall medications after the discovery of adverse health effects. A car OEM may recall models with defective or dangerous parts. The stakes are extremely high: in extreme cases, these defective items may cause injury or death.

As it now stands, batch recall practice is costly and cumbersome. Many supply chain partners need to combine data from their ERP systems. A recall ripples through many partners and batches and the affected product is often hard to separate from safe ones. Usually, companies are forced to err on the side of caution and recall many more products than needed. All this makes recalls very expensive.

Traceability is more difficult because partners may mix and match suppliers. For example, a bread supply line may include three different wheat growers, three different flour mills and three different bakeries—all of which may buy and sell to any of the others. In the event of a recall, how can the RFID system untangle the life history of a loaf of bread in this complex supply line?

RFID can help quickly identify products subject to recall. RFID systems can store batch numbers from the parts or ingredients used in all manufacturing steps. By tagging individual items, RFID allows tracing information to be stored and updated on the product itself. This allows recalls to be targeted down to the individual item, instead of sweeping up every possible candidate. In addition, RFID technology facilitates data exchange along the entire supply chain, and producers need not depend on another entity to share data.

But these systems raise fears of industrial piracy. Competitors potentially could use this information to snoop on a company's entire

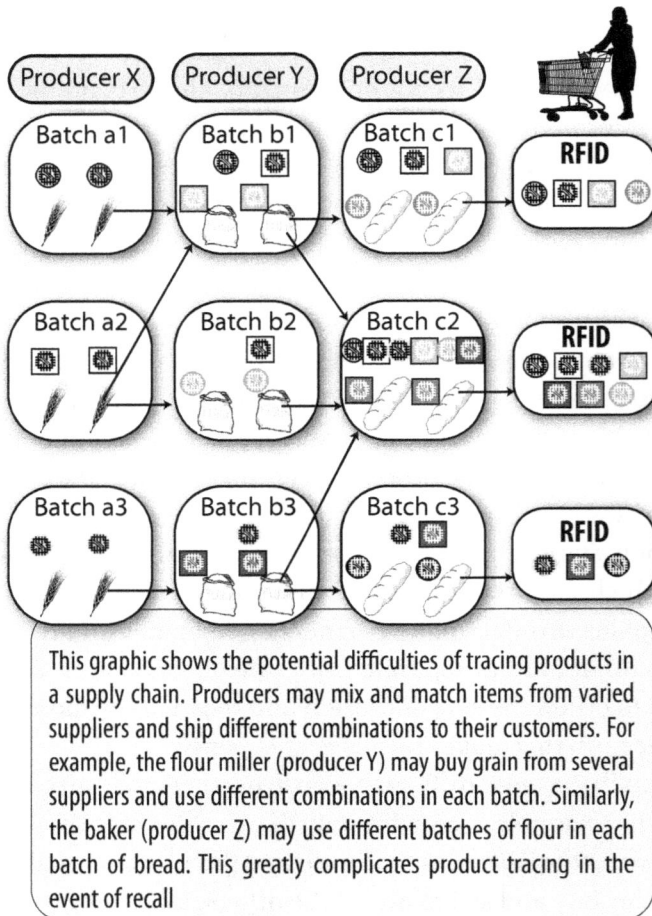

This graphic shows the potential difficulties of tracing products in a supply chain. Producers may mix and match items from varied suppliers and ship different combinations to their customers. For example, the flour miller (producer Y) may buy grain from several suppliers and use different combinations in each batch. Similarly, the baker (producer Z) may use different batches of flour in each batch of bread. This greatly complicates product tracing in the event of recall

*Figure 10-2. Supply Chain Example*

supply chain. For example, competitors could use RFID batch numbers to figure out another company's production volumes or supply chain structure. SAP researchers Leonardo Weiss Ferreira Chaves and Florian Kerschbaum explore this problem and propose solutions in the paper "Industrial Privacy in RFID-based Batch Recalls."[5]

[5] Chaves, L. W. and Kerschbaum, F. 2008, "Industrial Privacy in RFID-based Batch Recalls," in *Proceedings of the 2008 12th Enterprise Distributed Object Computing Conference Workshops* (September 16–16, 2008), EDOCW. IEEE Computer Society, Washington, DC, 192–198, DOI= *http://dx.doi.org/10.1109/EDOCW.2008.3*

Encrypting the batch numbers using traditional symmetric or asymmetric cryptography is possible but not practical. The producer would need to generate and store cryptographic keys for each batch and each company would need to be assigned a unique key space.

The SAP researchers identified one potential solution: storing tracing information on RFID tags and encrypting the information. According to these researchers, a solution should have three features:

- **Industrial Privacy:** All non-recalled batches should not reveal any information about the supply chain

- **Autonomy:** Any company should be able to recall all products that contain one of its batches without the help of companies downstream in the supply chain

- **Simple Key Management:** A producer should not have to maintain a key for each batch he produces

The SAP researchers developed a solution that addresses all three of these elements:

- **Industrial Privacy:** They present a universally re-encryptable identity-based encryption scheme that can re-randomize each ciphertext

- **Autonomy:** Each ciphertext is stored on the RFID tag and all are accessible to any consumer. Each company can independently issue recalls

- **Simple Key Management:** The solution uses the batch number as a key in an identity-based encryption scheme. No one has to maintain any key information, except the trusted third party

How would this work in practice? Each producer would attach RFID tags with product-tracing information and identity-based encryption. At each stage of the supply line, partners would add information about the products and batches they produced, again using identity-based encryption. Each producer would use an ERP system with a database that stores information about batches and private cryptographic keys.

Say a farm produces a batch of grain. Information about the batch is encrypted, written to an RFID tag, and sent to a grain mill to be ground into flour. The flour mill stores information about the batch and the private cryptographic information in its ERP system (each ERP system only contains the data on their own production batches). It is impossible to infer the batches used in the downstream or upstream supply chain because data is encrypted or unavailable. The flour mill uses different batches from the farm, each with an RFID tag with encrypted information about the batch. These are sent to the baker and the process repeats.

How to prevent people from inferring information about the supply line? How can the miller be confident that the baker won't be able to reconstruct information about his supply line? In this model, downstream producers re-randomize the ciphertext. They do not need the batch numbers and, without the decryption key, no one can tell that they are from the same plaintext. The encrypted information also is re-randomized. In other words, the miller modifies the encrypted information from the farmer without knowing his batch information. Each batch of grain will carry encrypted information from the farmer that looks different as long as the decryption key remains unknown. In the next step, the baker applies similar techniques. These processes repeat downstream all the way to the customer.

In the event of a recall, producers reveal the private cryptographic information about the bad batch to a trusted third party. The trusted third party alerts all retailers to remove the affected products and alerts consumers, who can check their own purchases. Someday, consumers will even do this with a smart fridge equipped with an RFID reader!

This method could allow batch recalls using RFID tags with universally re-encryptable identity-based encryption. As the authors conclude, "Our solution implements full industrial privacy, overcoming the main obstacle to RFID adoption in batch recalls."

These solutions suggest ways that supply chain management can become more accurate, trusted, and secure. No doubt, much work remains to be done on these technologies.

But Dr. Lotz is soberly realistic about how much security is attainable. As loops become more open and partnerships grow more dynamic,

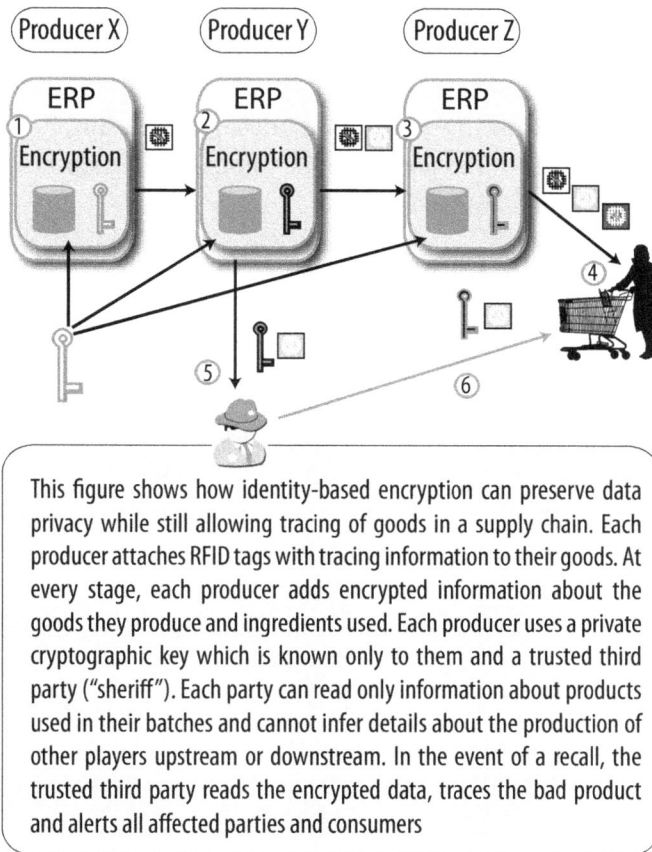

*Figure 10-3. Process for Using Identity-Based
Encryption in the Supply Chain*

enterprises may have no choice but to give up some control. For example, businesses increasingly outsource their computing and infrastructure to cloud providers and will have to accept whatever level of security the provider is willing to guarantee. Given that relations will be short and dynamic, it is not practical or feasible to renegotiate these arrangements in every case. In fact, Dr. Lotz is not sure that complete security is attainable in the IOT.

"If we look at the Future Internet, which comprises the Internet of Things and the Internet of Services and billions of largely distributed entities containing computing and application environments, I think we have to accept the fact that we will not be in a position to make it a secure and safe place," he said.

"So the challenge instead is to make a safe place for those entities that you are currently working with," Dr. Lotz said, "and simply accept the fact that complete security for the whole Internet of Things is something that we will never be able to achieve."

## Manufacturing

Manufacturing is fertile ground for the IOT. Factories tend to be closed loops almost by definition. Therefore, the unresolved problems like standards, integration, and cost-sharing pose less of a barrier in manufacturing than other sectors.

In the manufacturing world, IOT technologies already have been successfully deployed for tasks like asset tracking or logistics. Often, one company controls the entire process and is able to see a positive return on investment in a short period of time.

For example, Prof. Günther of Humboldt University describes one casting shapes manufacturer that began using RFID tags to locate equipment. Until then, workers spent a lot of time trying to find tools and other equipment on the shop floor. With RFID, they were able to pinpoint the items immediately and precisely. The company increased efficiency by 10 to 20%, which immediately paid for the technology.

Just as the manufacturing world has been an early leader in deploying the IOT thus far, it also may prove to be a leader in propelling it forward. The manufacturing world already has successful use cases. It is used to integrating partners and processes—just as vehicle OEMs put together cars composed of parts from many different suppliers. Moreover, the factory is the birthplace of most products. If tagging becomes commonplace, these tags may remain with the product until the end of its life. The factory is the birthplace of product memory and so these innovations can enable new innovations further downstream. Therefore, the manufacturing world is well positioned to propel RFID and other IOT technologies beyond limited closed loops.

"During the manufacturing process, the seeds for product memory are laid down," says Prof. Wahlster. "We want to enter all the information as a starting point for the diary of this individual product."

## Driving Forward

We are now seeing many interesting and mature applications of RFID. Current uses of RFID in manufacturing include:

- Asset tracking

- Asset use and reuse

- Production execution and quality control

- Product genealogy

- Maintenance, repair and overhaul (MRO)

- Inventory tracking and visibility

- Monitoring labor

For example, BMW has launched an RFID real-time location system (RTLS) at its assembly plant in Regensburg, Germany. According to *RFID Journal*, this system customizes each car based on its vehicle identification number (VIN) and automatically directs workers to the correct tool for the job.[6]

BMW's production line emphasizes customization. Buyers can specify specific interiors, seats, and engine parts when they order their cars. This customization must be done under strict time constraints: each assembly line station has just 50 seconds to complete its job. Each car requires specific parts and tools, and teams need instant information about each job. The assembly line produces about 1,000 cars per day.

The company had tried various methods, like barcodes and infrared technologies. But they required operators to put down their tools, pick up a barcode scanner, and then pick up their tools again for the assigned task. This process proved too time-consuming and prone to error. At times, operators made mistakes or failed to read the barcodes, and cars had to be sent back for correction.

The automaker partnered with Ubisense to deploy an ultra-wideband (UWB) RFID solution. The BMW Tool Assistance System combines

---

[6] Swedberg, Clare, "BMW Finds the Right Tool," *RFID Journal*, Aug 4, 2009, *http://www.rfidjournal.com/article/view/5104*

Ubisense RTLS technology with IBS tool-controlling software. Deployed in January 2009, this system allows the factory to track assets, vehicles, and tools at 120 stations on the busy assembly line.

This system allows for extreme precision. The RTLS tracks the whereabouts of each item within 15 centimeters—not bad considering the assembly line is 2 kilometers long! Vehicles are separated by only about one foot of space. At each station, as many as five tools may be used.

At the start of the production line, a worker encodes the car's VIN into an RFID tag and places it on the hood. As the car moves down the line, 380 readers keep tabs on all RFID tags in the vicinity. At workstations, each tool also bears an RFID tag with a unique ID number. This system not only identifies each tool, but also whether it is in use or idle.

Back-end systems keep tabs on the location of each car and direct workers at each station to install specific parts or use specific tools, based on the customer's order. At the end of the line, the quality control team checks each vehicle. If it passes inspection, the tag is removed and returned to the start of the line for reuse.

This example shows how RFID and other IOT technologies can transform the shop floor. But it also exemplifies the closed nature of most existing manufacturing applications. How can these systems become more open and involve more players?

## Prognosis

According to Mr. Neidecker-Lutz, manufacturing companies are well-positioned to push the IOT beyond closed loops. Not only is the manufacturing world an early adopter, but also it is used to mingle different partners and processes. "A lot of the issues that are problematic in the general case are not really a problem in the manufacturing space," he says.

He sees two main pieces: the factory and the product itself.

The first aspect—the factory—is relatively straightforward. Often there is one owner so there are fewer questions of who pays and who benefits. They are closed environments, where the ownership and technological challenges are clearer. Granted, the shop floor can be a hostile environment for technology, with a lot of cacophony, activity, machines, and electronic interference. But many of these problems have been overcome and deploying simple edge devices has become a fairly simple matter. The challenge grows as these systems grow more complex, with full-fledged computers on the edges and more complex software behind them.

"The frontier there is in how much complexity in configuration management, information exchange can you manage before you can no longer cope?" asks Mr. Neidecker-Lutz. "That's sort of the limiting factor today for these kinds of environments."

"You can sort of do this within any one company," he adds. "But since manufacturing has gone very much cross boundaries, the challenge is in how to consistently exchange information across the chain. The charge is being led not so much by the communication portions, but rather by putting information into the things being manufactured, because they naturally already travel at the right time across the right kinds of company boundaries. Everything past the early design stages can be augmented by local product memory."

The second challenge—putting technologies on the product itself—is much more complex. It raises many issues discussed elsewhere in this book: standards, integration, data ownership, value-added services, privacy, security, digital product memory, business models, and so on.

Manufacturing already has taken steps in this direction. For example, OEM car companies might make only half the components of their own cars and outsource the rest. "For a lot of the discrete manufacturing industries, you have pretty deep hierarchies of different companies working together, and you have distributed control, distributed configuration management for these things," he says.

Manufacturing chains are well positioned to develop new business models that take advantage of the IOT. Once again, the question is, how can we build value chains that not only extend the infrastructure of the IOT to many players, but also allow each of them to extract value from it?

## The Next Wave

Rick Bullotta of Burning Sky Software believes that manufacturing is a bellwether that will lead the way for the deployment of the IOT. Factory environments are closed-loop, mission-critical applications with high stakes, and thus have been pressured to develop solutions to key questions on the IOT.

"There are a lot of difficult problems that have been solved in that domain," he said. "There are probably a lot of technical learnings that can be brought over to the general connected world, Internet of Things concept."

Mr. Bullotta expects factories to undergo a wave of progressive automation. Machines will increasingly replace humans in performing tasks. The IOT will move from sensors and actuators toward artificial intelligence and decision-making. Even the wrench-wielding shop floor laborer will become more of an information worker. Processes will become more adaptable and more ad hoc. As a result, companies will be able to exploit niche opportunities of the long tail.

He is bullish about the IOT but not about the devices *per se*. Instead, he is excited about how it enables collaboration between people, devices, and systems. It can enable work processes through real-time intelligence with more applications that are truly connected to the physical world. It allows us to work in an environment where decision-making and action are increasingly compressed. It opens the door to real-world awareness.

Mr. Bullotta envisions "disposable companies." The founder might serve as the brand owner and recruit players for all the necessary processes: a design firm, manufacturer, marketer, distributor, and so on. "All the pieces of that chain are getting broken apart," he said. "That creates a whole new dynamic that's needed for that real-time communication and coordination of the information from that product and about that product."

The IOT opens the door to a new form of product life memory that captures every aspect of the processes from design to disposal. Each facet can be extracted, analyzed, replaced, or repurposed. According to Mr. Bullotta, this opens up an array of new use cases.

As he noted in his IRF presentation, the manufacturing world may lead the way for new business models. For example, a product like a car may become not just a product to be assembled, sold, and forgotten. Instead, it becomes a platform upon which to offer future products and services. The factory becomes the beginning of a whole new relationship with the consumer.

## Retail

The retail sector has been a leader in applying RFID. Wal-Mart and Target in the US, Tesco in the UK, and Metro in Germany all have rolled out major RFID initiatives.

Retail is a crucial frontier of the IOT. The efficiencies that the IOT can bring to manufacturing and logistics are meaningless without good execution in the last few yards of the supply chain.

RFID is more than just a substitute for the barcode. It brings automated stock management and, potentially, high-resolution management to the retail floor. One example is out-of-stock (OOS), which is a persistent shortcoming for most retail businesses. Most retailers report OOS levels of 5 to 10%. Nearly three quarters of these shortages directly result from retail practices such as inaccurate forecasting and ordering or even having the product in the storeroom but not on the shelf. With RFID, stores can keep tabs on shelf stocks in real time.

At this point, most companies tag only larger quantities of goods such as containers, pallets, or cases. In the future, they are most likely to introduce RFID to increase data quality where the benefits of higher data quality exceed the cost of installing these devices. They are also most likely to tag items of high value or those that suffer from high rates of shrinkage, or depletion due to loss or theft (condoms and razors seem to be items that frequently vanish from inventories).

We can catch a glimpse of this world in the Future Retail Center in Regensdorf, Switzerland. This project is run by SAP Research in cooperation with partners such as Migros, Siemens, HP, and Nokia. The store is refining end-to-end processes that integrate SAP ERP and retail back-end systems and Auto-ID infrastructure. It is the first such living lab in the world that showcases how the IOT may transform the retail sector.

Products are tagged with RFID instead of conventional price tags. The store shelves are equipped with RFID readers to keep track of inventories. SAP retail back-end systems keep track of stocks, receipts, and sales.

An RFID reader in the bottom of the cart logs each item picked by the shopper and displays it on a screen mounted on the cart. When the customer picks out fruits and vegetables and places them on the scale, the system prints out a tag with weight and price.

A "quick shopping" mobile retail scenario allows customers to store their shopping lists on their smart phone. When they walk into the store, their list is automatically matched with the inventory. The system will even help them navigate the store and guide them to these items with a visual map on their phones.

Similarly, other applications may suggest items they might like, based on personal preferences, or offer electronic coupons.

At checkout, the customer simply has to pass the shopping bag past an RFID reader that instantly totals the items. The customer can pay with a credit card or mobile phone. Customers simply lay their cards or Near Field Communication (NFC) mobile phones on the scanner, enter their PIN numbers, and checkout is done. No more waiting for slow cashiers in long lines.

Meanwhile, in-store systems keep tabs on shoppers. When a customer passes with the cart, the screen might flash advertisements of particular interest to that shopper.

The Future Retail Center also explores scenarios for logistics and retail strategy. The system alerts store operators when items need restocking. In the storeroom, all items are tagged with RFID. A forklift is equipped with an RFID reader and computer screen and can guide the operator to pallets that need to be moved.

For those who can't visit in person, take heart: the Future Retail Center has a virtual twin in Second Life where customers can shop with their avatars.

## The Internet of Services

As the IOT matures into more open loops, it merges into the realm of services. The IOT must grow hand-in-hand with the Internet of Services. In fact, services may be yet another killer application of the IOT.

In the future, companies will increasingly expose data to partners and customers through the Internet of Services. One example is delivery companies like UPS and DHL, which developed sophisticated barcode systems to track shipments. These systems were developed for internal use but later shared with customers. As a result, you can now go to the company web page, enter your tracking number, and check the location of your package.

"I think there's a huge potential in bringing the Internet of Things and Services together," said Dr. Uwe Kubach, SAP Research. "Just expose the data that you gather through the Internet of Things through the Internet of Services, offer this to your customers in an easy, consumable way so they just go to your web site, access the service, and have that information as well."

Another example is the smart car. Our personal cars may serve as platforms that access both the IOT and Internet of Services. Car dashboards might serve as menus for services. They will tell us where we can find gasoline, lodging, or food. They may alert us when a part is in need of replacement and offer a menu of possible vendors of services to fix it. We can already see fledgling examples of these trends: dashboard GPS systems, turn-by-turn-directions, and on-board computers. Indeed, electronics are expected to soon account for 20 to 30% of the cost of an average vehicle. Prof. Wahlster envisions a future where these smart cars are equipped with semantic technologies and allow the driver to say, "where can I find the cheapest diesel?" or "what's playing at the movies?" and have the car answer with a list of options.

Smart cars represent one example of how the IOT and Internet of Services will become so integrated that it becomes difficult to distinguish between the two.

### Turning Data into Services

Imagine the plastic turning knob of your washing machine breaks and you need a new one. You pull out your smart phone and read the RFID tag attached to the machine to identify the model number. The display flashes a menu of services: instruction manual, warranty information, and spare parts ordering. The phone could scan the part to identify it and then present several places that sell it.

"So suddenly you're creating a service market," says Dr. Rode. "This is not about the Internet of Things; it is an Internet of Services topic."

Dr. Rode believes that these kinds of service offerings represent a potential killer application of the future. This scenario could apply to almost any product in the world. These services not only could supply parts, but also contact repair technicians, schedule maintenance, and suggest when to replace your obsolete machine.

This levels the playing field and allows smaller companies to compete with larger ones that formerly had a vertical monopoly. For example, a small company that specializes in plastic turning knobs could compete for your business alongside the OEM that made the machine.

## What Is the Internet of Services?

The Internet of Services transforms the classical Internet into a service marketplace. It combines services of the classical Internet with newly automated ones and allows for easy consumption, easy mashup, and easy delivery through the Web in a service-oriented manner. It forms the basis of an emerging global infrastructure that will give rise to more flexible business networks and value-added services.

The Internet of Services enables companies to be more agile, focus on core competencies quickly, form ad-hoc partnerships, and extend their reach into new global markets. The future service platform rests on four cornerstones: the web-based service industry, cloud infrastructure, future of killer applications, and semantic service discovery.

The IRF 2008 was devoted solely to the topic of the Internet of Services. Here is a summary of these trends distilled from the previous IRF book.

### The Web-Based Service Industry

The Web is revolutionizing the service industry. Thanks to advances in the service grid and communications technology, barriers have fallen and connectivity and computing are nearly ubiquitous. Software-as-a-service, value-added services, the growth of cloud infrastructure, and advertiser-subsidized business models have opened new possibilities for offering services on the Web. As a result, we are witnessing the emergence of a new web-based service industry.

Services are becoming composable and tradable over the Internet just like real goods. In the future, we will be able to mix and match services and compose business solutions with the same ease that we assemble playlists on iTunes. In this new service industry, both startups and established players can compete on a level playing field.

We are beginning to see the emergence of marketplaces and aggregators that gather, organize, integrate, and broker these services. We can see models such as SAP's TEXO project and, increasingly, real-life examples in the world of businesses.

### Cloud Computing

The cloud model represents a revolution in computing infrastructure. The cloud model allows companies to outsource basic computing needs and purchase information technology as a service. The cheap availability of cloud services opens the door to potential innovators who otherwise might face substantial barriers from capital costs if they had to invest in a data center. This democratizes innovation and gives small startups access to unlimited resources at low cost. Rather than worrying about keeping the computer engines running, they can focus on their core business.

### Killer Applications

It is impossible to develop a simple formula for the killer app. Indeed, these successful applications often seem brilliant only in retrospect. At first, they often seem crazy or pointless. Yet the IRF 2008 identified several key traits of successful products: they must be simple, easy to use, and highly scalable. They must solve the "pain points" of a business problem or bring joy to their users. Experts identified some key areas ripe for development of killer apps—particularly mobile technology. They also identified a number of potential killer apps that rely heavily on the IOT, such as smart buildings for energy conservation and assisted living.

### Semantic Service Discovery

Service discovery is the *sine qua non* of the web-based service industry. Yet it remains an elusive quest: current service discovery techniques rely on keyword, metadata, and ontology-based search. This approach is suitable within enclosed domains but is not viable in the global Internet of Services. We need more effective methods for finding and advertising services.

Here, the Internet of Services runs into the vision of the Semantic Web. The Semantic Web is a vision for a next-generation Web in which technologies and standards express meaning of data so that machines can understand it.

If this vision can be attained, it opens a new world of possibilities for automation and service discovery. Yet this vision remains dogged by shortcomings and many critics see it as a "pie in the sky" fantasy.

The infrastructure barriers to doing business in the Internet of Services are dropping rapidly. These trends are democratizing innovation as never before. At the IRF 2008, Dr. Neel Sundaresan of eBay envisioned a day when an entrepreneur will be able to launch and run a business— managing everything from raw materials, to production, to transport, to point of sale—almost entirely from a mobile device.

*Integration*

The Internet of Services depends on the ability to freely combine multiple independent businesses. Devices and IOT networks must be able to talk to each other. This requires clearer standards and, perhaps, semantic interoperability. SAP Research and its partners recently made progress toward one important aspect of this challenge with the creation of a new unified services description language (USDL). This language allows services to be exposed beyond company "firewalls," on the open Web. USDL represents an improvement over web service description language (WSDL) and offers a richer description of features such as ownership, availability, pricing, and technical aspects. As such, it represents an important step toward a vision of a future of easier service discovery, interoperability, and the flexibility to repurpose services through new channels and applications.

## Anti-Counterfeiting

Pirates not only sail the high seas; they also surf the Internet. The technological revolution has unleashed new waves of piracy and counterfeiting. The IOT may offer tools that help deter dangerous counterfeits and cheap knockoffs.

According to the International Chamber of Commerce, an estimated 7% of global trade is counterfeit goods. Consider these statistics compiled by Prof. Fleisch and his ETH Zürich and University of St. Gallen colleagues Thorsten Staake and Frédéric Thiesse:

- Nearly half of all motion picture videos, more than 40% of all business software, and a third of all music recordings are pirated copies

- About 10% of clothing, fashion, and sportswear are counterfeits

- In the automotive industry, 5 to 10% of all spare parts are counterfeits

- Up to 12% of toys sold in the US are counterfeits

- Between 5 and 8% of medicines sold worldwide are counterfeit, according to estimates of the Word Health Organization. In some developing countries, the counterfeiting of drugs is endemic. In 2001, 192,000 people died in China due to fake drugs

Piracy and counterfeiting damages many parties. Consumers may save a few bucks, but have fewer protections and may even put their lives in danger with unsafe products such as drugs. EU Commissioner of Enterprise and Industry Günter Verheugen recently warned that "every faked drug is a potential massacre." Governments are unable to regulate or tax these black market goods. Legitimate owners of brands and intellectual property are robbed of their profits, and may even be blamed when bogus goods cause problems. These enterprises can face unjustified liability claims, damage to brand image, loss of revenue, and a decreased return on investment from research and development.

In some developing nations, piracy represents a major proportion of the economy. The profits of counterfeiting may even exceed drug trafficking. Yet the risks and potential penalties are much smaller. One kilo of pirated CDs is worth more than a kilo of cannabis resin. In many European nations, selling counterfeit goods might bring a 2-year prison term and a €150,000 fine. The same person convicted of selling drugs might face 10-years in prison and €7,500,000 fine. Is it any wonder that organized crime is turning to piracy?

Many companies already take anti-counterfeiting measures, like holograms or elaborate hard-to-duplicate packaging, biotechnology, or microelectronics. Unfortunately, these techniques are not suited to automation warehouses that see vast volumes of goods coming and going each day. Nor do they provide adequate security.

## The Potential of RFID

RFID, and perhaps future IOT technologies, provides a promising alternative against counterfeiting and piracy. It becomes possible for companies, customs authorities, regulators, and consumers to track and trace the life history of a single item. This visibility makes it easier to detect illicit goods in a supply chain and deters piracy.

For example, a shopper in the UK may use her mobile phone to check the pedigree of a drug package and be justifiably suspicious when she discovers that the batch is supposed to be in Nigeria at the same time.

The potential of RFID in anti-counterfeiting has been explored extensively by Prof. Fleisch and his University of St. Gallen colleagues, Thorsten Staake and Frédéric Thiesse. These researchers outlined how one model might work. A network like EPCglobal, or a similar network or service, would provide the IT infrastructure with the extension of local information, discovery, and authentication services. The channel itself does not have to be secure.

These methods could be used with simple RFID tags (which can be cloned) or, in the case or more valuable items, RFID tags with secure authentication. In cases of high-value items, such as airplane parts, more secure computing methods may be justified. The tag contains a unique identification number, a secret key, and a cryptographic unit. The associated key is stored on the database of the network or an encryption service. When it comes time for authentication, the tag communicates its identity number with the manufacturer's cryptographic unit (CU). The CU responds by sending a random message (challenge) back to the tag. The tag encrypts the message with its secret key and sends the response back to the CU. The CU looks up the associated key in a database and verifies the response.

These anti-counterfeiting platforms could be supplied by either a large network like EPCglobal, or a specialized platform. Until now, there has been no comprehensive, enterprise-grade service to combat piracy and illegal business.

### Envisioning Authenticity

One new company in this field is Original1, a global platform that certifies authenticity. Original1 tracks the supply chain from manufacturer to wholesaler to retailer. Users will be able to check the history of a product to verify its authenticity or expose counterfeits.

Original1 is a joint venture that combines the competencies of SAP, Nokia, and G&D. SAP back-end technology will follow products through the entire lifecycle of factory, wholesalers, shipping customs, retailers, and customers. Nokia technologies will provide mobile connectivity and authentication. The data will be hosted by Original1 and protected by technologies created by G&D. All these elements will be tied together in Original1's global SaaS platform.

Original1 seeks to create a brand-protection platform with end-to-end transparency. It rests on the idea that in the future authenticity will be a key driver for sustainable success for many companies.

Enterprises currently lack a structured approach to this problem. Counterfeiters apparently have some wily technologists on their side. Ms. Alsdorf of SAP Research, who now heads Original1, recounts a recent trip to China where she saw holograms on packages of Nokia cellphone batteries that had been faked.

"How secure is the data, and is there the possibility along the supply chain to change and manipulate the data, even within this kind of clearing house function of Original1?" asks Ms. Alsdorf. "People are very nervous about that. One technique we are trying is using encryption technology to make sure that nobody has unauthorized access to this kind of data or can manipulate it."

Original1 will allow mobile users to check the authenticity of a product at any point in the product lifecycle, from factory to consumer. This service allows easy and fast access anytime and anywhere. It uses encryption to ensure that sensitive data cannot be accessed by unauthorized users. It will integrate authentication technologies like RFID, counterfeit-proof printed labels, and machine-readable holograms.

It will feature an architecture for data protection and mobile interfaces that integrates enterprise back-end systems and mobile devices. It will feature packaging hierarchies (pallet, box item), product documentation with GS1 numbering schemes, and even non-standard formats customized to brand owners. The service will be adaptable to different use cases and data sources. It will be able to detect anomalies through serial tracking and tracing.

Each party will be able to check the authentication of the product at any point. For example, a person could use a cell phone camera to take a photograph of the security marking of a product and send it to GBPS (Global Brand Protection Service). Within moments, the cell phone would return a message verifying the product is indeed an original.

For example, a customer might query about a handbag in a retail store in Germany and discover that the product was supposedly shipped from China yesterday. "You know something is wrong because that cannot be," says Ms. Alsdorf. "If you have this kind of supply chain information, you can verify easily that this cannot be the product that was supposed to be shipped from China to Germany yesterday."

These technologies also can transform customs. According to Ms. Alsdorf, EU customs inspectors can physically check only 3 to 6% of shipments. These technologies could enable customs officials to be more thorough and target their efforts. There is ripe opportunity to involve consumers, especially in retail apparel and sportswear.

Original1 plans to begin with pilots in the pharmaceutical and chemical industries in Europe and eventually expand to other industries in Europe, Asia, and the Americas.

But Ms. Alsdorf predicts that these anti-counterfeiting platforms may take time to catch on. For the most part, companies do not bother to invest in them unless forced to by government regulation or legal liability.

"If the brand owners are not really suffering, they see no need to invest in it," she says. "Loss of revenue is not really an issue yet."

As Ms. Alsdorf notes, the technologies are ready to deter counterfeiting. RFID chips are small and durable enough to be sewn into tags and even put through the washing machine. The larger barriers are the business models and politics.

"The retailers ask, 'who's going to finance it?'" says Ms. Alsdorf. "It's a battle between the brand owner and the retailer. It's not yet clear who is going to own what and who's going to pay for what."

Once again, the issue returns to the political economy of the IOT. Who pays, and who benefits? It also raises the familiar questions of trust and security. According to Ms. Alsdorf, many companies are hesitant to share data. A retailer may be reluctant to share data with the clothing manufacturer. But, she adds, they are more apt to participate if there is a trusted third party.

## Other Solutions

In some sectors there is more urgency. Piracy is an acute problem in China. According to Prof. Min, some products may sell for lower prices in poorer provinces and higher prices in more affluent places, like Shanghai. Sometimes companies claim they will sell a product in a poorer region and really sell it in an affluent one. They lie to get higher margins. RFID can also help guarantee legitimacy in those cases, he says.

In one case, rampant piracy prompted one liquor company to install RFID tags on liquor bottles. Prof. Min helped develop an anti-counter-feiting system that allowed the company to guarantee the authenticity of each bottle.

The Wu Liang Ye Company is a major liquor distiller in China. The company is famous for its Wulling Liquor, a strong aromatic liquor made from broomcorn, rice, glutinous rice, wheat, and corn that is based on a 600-year-old recipe handed down from the Ming Dynasty era. The bottles sell for $100 each and the premium brand has brought counterfeiters hoping to siphon off some of the company's profits. According to Prof. Min, the counterfeit rate was estimated to exceed 50%. Sales plummeted because people were reluctant to shell out $100 for a product that had a 50-50 chance of being bogus.

"Because of counterfeiting, people hesitated to buy this kind of liquor," he said. "People start to buy a lower-priced kind because it had less counterfeiting."

The Wu Liang Ye Company approached Prof. Min and his colleagues at the university for help. They suggested a solution: put RFID tags on each bottle. They put together a team that included manufacturers of tags, readers, and systems integrators.

Every bottle will have an RFID tag inside the box. At every store there is a kiosk with an RFID reader where the customer can check the authenticity of the bottle in their hand. The screen will identify the original manufacturer, where and when it was made—all verified by a digital signature from the owner. The new system will go online early in 2010.

This RFID system guarantees that the bottle is authentic and thus adds tremendous value. At this point, the liquor is the only product using this network, but once it becomes established other products may piggyback on the proven system and join.

As Prof. Min says, the system allows the company to guarantee the authenticity of his product: "If you buy my product in the store and you check it with the reader, you know it is the real thing."

## Smart Energy

Energy efficiency and climate change are hot topics at the forefront of corporate and government agendas. Energy management is a fertile opportunity that has attracted the interest of companies like SAP, Google, and IBM.

The IOT has the potential to transform the model for providing utility services. In the future, homes, businesses, and factories may be equipped with sensors that monitor and adjust usage of electricity, gas, and water. Utilities will manage smart grids in which real-time data allows them to more efficiently balance loads, make greater use of renewables, create dynamic marketplaces, and offer new services.

As Dr. Harald Vogt, a researcher at SAP Research puts it, "These are basic technologies which are now being used to turn the whole energy supply upside down."

## Changing the Paradigm

Today electrical energy distribution follows the "broadcasting model" with a few large, monolithic centralized power plants that serve millions of users. In the future, experts predict a shift toward a community-based model with many decentralized power systems.

"One of the major challenges is to get timely and correct information about energy consumption and production in the grid," says Dr. Vogt. "This information is not available today. The grid operators have an overall picture, but they don't have a fine-grained view into the grid and into all the devices attached to the grid."

**Dr. Harald Vogt** has been a researcher at SAP Research since 2006. His main research interests are distributed systems, pervasive computing, and software engineering. Currently, he is active in developing systems for the operation of smart electricity grids and electric vehicles. His results will help to prepare SAP's industry solutions to meet future requirements in the utility sector. Before joining SAP and pursuing his Ph.D. in secure communications for wireless sensor networks, he spent a few years at the Darmstadt research center of Deutsche Telekom AG, where he investigated the security of Java smartcards. Dr. Vogt holds a master's degree in computer science from the University of Ulm, Germany, and a Ph.D. in computer science from ETH Zürich.

Enter the IOT.

New technologies can create smarter systems for metering buildings and new infrastructures for producing and distributing energy. This field goes by several names: the smart grid, Internet of Energy, or e-energy. All convey the same goal: to make more intelligent use of energy, improve efficiency, and reduce greenhouse gas emissions.

This could bring a paradigm shift in how we produce and consume energy. It could introduce new business models. Until now, long-term utility contracts have been the norm; based on this new information, short-term contracts might become feasible.

"This whole market environment is likely to change," says Dr. Vogt, who researches how to create more intelligent, data-driven models for energy production, management, and consumption, "With these electronic metering devices, new things will be possible."

Consider a few examples.

Utilities or customers could instantly call up an account's current usage. Utilities could reconnect an account with a software transaction instead of sending out a technician to the site. Utilities could use advanced metering infrastructure to share pricing information with a customer's smart meter, enabling a local energy management system to optimize consumption.

Consumers might adjust their daily energy use and secure better rates by conservation. A customer might schedule charging an electric car at the optimal time. A customer might agree to automatically shut off certain appliances when power loads become too high.

This data also will bring new models for services, just as the Internet already has done in many other sectors. The system might detect that your refrigerator is obsolete, triggering an alert and advertisements for newer energy-efficient models. It might warn a customer that his water bill is much higher than usual, suggesting that there may be leaky plumbing. Once again, the IOT leads to the Internet of Services.

"As a utility, I can more closely watch the usage behavior of my customers and maybe come up with new contracts or other contract offerings to them," says Dr. Gregor Hackenbroich of SAP Research. "That adds to customer satisfaction."

A smart grid also transforms the supply side. Automated systems will help manage complex tasks, such as load balancing, reconciliation of traded energy volumes, and determining revenues, costs, and fees. Utilities would be able to immediately pinpoint areas affected by power failures. When an overload is imminent, pre-programmed systems could instruct meters to switch off appliances.

Internet marketplaces will allow private producers to trade energy under contracts. These marketplaces could make independent, entrepreneurial energy producers more viable. These distributed energy sources would diversify the energy supply. It could spur green energy. Renewables like wind and solar power have one major shortcoming: they are intermittent and therefore have to be balanced with conventional sources of energy. With a more data-driven and dynamic market, utilities could enlist customers in helping to balance supply and demand. Citizens might have the option to buy only green energy from solar or wind producers. By making clean sources of energy more viable, the IOT could help reduce environmental impacts, such as the emission of greenhouse gases.

Someday electrical car drivers might enjoy "energy roaming" privileges for their electric cars just as they do for their phones. Charging stations will enable people to plug in anywhere on the European continent and have the costs added to their electric bill.

## Realizing the Vision

So that is the vision. How to get there?

Klaus Heimann, SVP of Service Industries for SAP, emphasizes one vital message: successful applications must be useful to customers. They can't just focus on data management or technology. This harkens back to our earlier discussions about political economy, sharing benefits, and developing new business models.

"At the end of the day, smart metering was made for consumers for the purpose of making them more energy efficient" says Mr. Heimann. "In order to roll out the smart grid, you have to get the buy-in from the consumer—and, by the way, that can go terribly wrong."

Unfortunately, he says, many technicians don't realize that their technical zeal is not shared by the average consumer. "They forget about making this technology attractive, usable, and affordable to the consumer," he says. "They just love to talk technologies and they forget about why we are doing this. And very often they come up with proposals where the private consumer just shakes his head, thinking, 'Man, do these people not have proper work to do?'"

## A New Solution

SAP designed the utilization of smart metering systems with consumers foremost in mind. It piggybacks on the company's existing relationship with utilities. SAP's Customer Relationship and Billing solution serves about 700 million private, commercial, or industrial customers, about 40% of supply contracts worldwide for water, electricity, and gas. About 750 utilities worldwide use this SAP solution.

"We really focused on how we can bring all this technology to the consumer," said Mr. Heimann. "How can the consumer actually get his arms around that technology in a way that he understands and is beneficial to him? If it's not beneficial to him, he will simply not do anything."

According to Mr. Heimann, we can think of the smart grid as a nerve system that monitors and controls the power supply chain. This includes power from all sources of power generation (supply side), transmission and distribution through the power grid, and any energy-consuming asset/appliance of the consumer (demand side). This powerful IT network is often called the Internet of Energy and includes very precise

**Klaus Heimann,** Senior Vice President within SAP AG's Business Unit Global Industry Solutions, is responsible for SAP's software solutions for service industries (media, telco, utilities, and waste and recycling).

From January 2002 through April 2007, Mr. Heimann headed up the Industry Business Unit (IBU) Utilities at the headquarters of SAP AG in Walldorf and was responsible for SAP for Utilities, SAP's market-leading business platform for the global utilities industry.

Over his 32 years of experience in information technology, Mr. Heimann has concentrated on the development and implementation of standard software solutions for various service industries, particularly for the utilities industry. Before joining SAP, he worked for 16 years at a German software company, where, as General Manager and a co-owner, he focused on building standard software solutions in customer relations and billing for the European utilities industry.

Mr. Heimann studied information technology at the University of Karlsruhe, Germany, graduating with a Diplom-Informatiker.

rules and restrictions about write access or even read access. According to Mr. Heimann, SAP made an important achievement in the support of its utility customers and partners—integrating the smart grid (in particular its demand side) to SAP's Customer Relationship and Billing solution. "This integration is mandatory to make the smart grid usable for and accessible to the consumer," says Mr. Heimann.

This IT network would consist of several layers:

- Local devices such as home appliances or generation units
- Home automation systems such as on-premise thermostats
- Advanced Metering Infrastructure (AMI) such as smart meters, concentrators, and data collection systems
- Meter data management systems that act as the central hub
- Central applications, such as software for CRM billing and management systems for data, assets, and outages

How to move toward this model? SAP formed the AMI Lighthouse Council in North America and worked alongside utility companies to identify top priorities. The Lighthouse Council developed a model for how data management systems should work and identified key functional enhancements:

- **On-demand meter read:** The ability for a utility to instantly look up a customer's current usage

- **Utility reconnects customer:** The ability to instantly and remotely recommend service in case of a delinquent account or connecting new service

- **Price signal:** Utilities could use the AMI system to share pricing information, enabling the consumer or his energy management system to optimize electricity consumption

In this model, the meter data management software acts as a gateway to the AMI systems and all the smart meters. SAP for Utilities acts

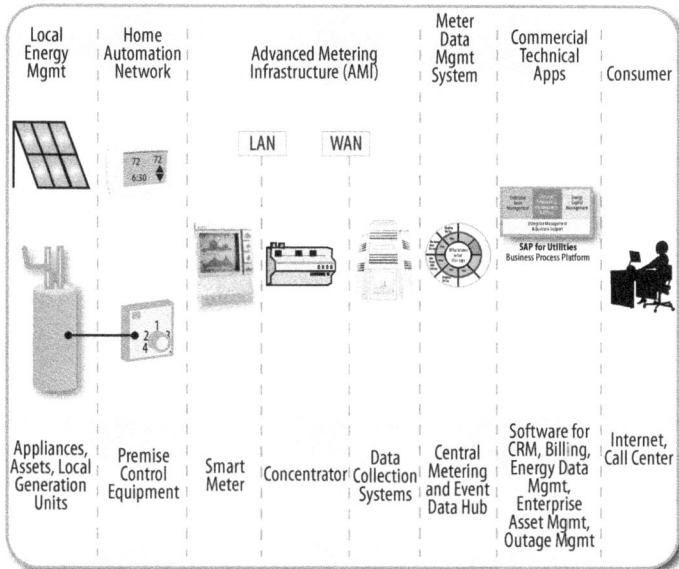

*Figure 10-4. The AMI Enterprise System Structure*

as the central hub. This model consists of Meter Data, Unification, and Synchronization (MDUS):

- **Meter Data:** The MDUS is the system of record for meter readings and other data. It receives information directly from AMI data collection systems

- **Unification:** The MDUS integrates with AMI systems from various manufacturers. It makes these different systems transparent to back-end systems

- **Synchronization:** The MDUS will obtain additional data from back-end systems, such as SAP for Utilities and synchronize itself with the AMI systems. In turn, each AMI system will synchronize to its meters

Based on this input, SAP released the SAP AMI Integration for Utilities software in late 2008. This software is based on SAP ERP, enhancement package 4, and SAP Customer Relationship Management (SAP CRM) 7.0. It also includes services that allow the MDUS systems to communicate with SAP for Utilities.

## The Internet of Energy Needs the Web

The Internet of Energy needs the Web in many ways that are too detailed to explain here. Mr. Heimann illustrates with a brief example that he expects to soon emerge in the market.

Making consumers more energy efficient requires some cooperation between consumers and their local utilities or, in liberalized markets, their energy retailer, simply because the consumer has more options and information to consider. According to Mr. Heimann, contacting the call center of a utility is not the best communication channel for this cooperation. A call center is expensive for the utility, and at the end of the day the consumer pays for those expenses. Call centers also have very limited ability to describe products or services and tend to be closed when the consumer has to call it.

Web-based customer self-services would be a much better communication channel to facilitate this cooperation. Mr. Heimann outlined a vision for such a customer self service platform with some examples:

- Visualizing the consumption behavior of the consumer by turning the 15-min KWh values into an easy-to-read graphic

- Graphs showing how many KWhs have been generated by the solar panels on the customer's roof top

- Detailed explanations of monthly or quarterly bills that review every line item and suggest how the consumer could potentially reduce costs of this line item

- A self-service tool that allows consumers to build models of their homes and drag and drop in symbols for their energy-consuming appliances. This tool would allow consumers to easily build their own energy management systems. It would answer questions such as: What are the energy costs of specific appliances? Which appliances are worthwhile to replace? How would my bill change if I switched to a new plan?

These kinds of customer self-services would reduce call volumes at the utilities' customer service centers and would generate savings, helping the utility to fund the huge investments required to build the smart grid infrastructure.

## Research Frontiers

We already can see early signs of other new models emerging. There are wholesale markets for energy, but these are limited to big suppliers and not accessible by small producers or consumers. There are smart grid projects in the US, Australia, and Europe.

Germany is a leader in research. One example is the e-Energy program of the German Federal Ministry of Economics and Technology. The projects within this program have been set up to demonstrate how information and communication technologies can achieve greater efficiency, supply security, and reduce the environmental impact. The program uses regional model projects to test how IOT systems may enhance and optimize the energy supply chain.

Another example is the MEREGIO (Minimum Emission Region) project, which seeks to optimize regional power systems to reduce greenhouse gases. SAP is working with partners such as Energie Baden-Württemberg AG, ABB AG, IBM, Systemplan GmbH, and the KIT (Karlsruhe Institute of Technology). MEREGIO integrates an e-energy marketplace, an innovative energy infrastructure, and a powerful information infrastructure. The project is developing a 1,000-customer pilot project in the Karlsruhe/ Stuttgart area of Germany and standards that encourage other regions to reduce greenhouse gas emissions.

On the European level, SAP Research is a partner in the Smart House/ Smart Grid research project. This project examines how the IOT may create a better model for better sustainability and energy efficiency. The project is conducting field tests in three European countries. In the Netherlands, they are testing large-scale communication systems that link thousands of smart devices. In Germany, another part of the project examines how to intelligently interact with customers and optimize home energy management. In Greece, they are examining how to optimize aggregate energy efficiency while ensuring supply security.

Many challenges lie ahead. In the future, real-time systems will have to contend with a deluge of data. Instead of reading meters once per month, they will do so every 15 minutes. "About 100 values per day per meter, and we're talking about millions of meters, or even more, billions maybe in the future," says Dr. Vogt. "This is really a lot of data that has to be managed."

How can we create new billing formulas for both power consumption and access to the grid itself?

## Summing up

These challenges will be met one step at a time.

"From a vision perspective, you have the feeling it's a paradigm shift and it's a revolution," says Mr. Heimann. "But things are not eaten as hot as they are cooked."

In other words, he cautions that this shift will not occur overnight. Companies must be careful not to roll out technologies and systems that are not ready to scale up. And they should carefully define standards so they don't have to spend a lot of money later trying to make systems compatible.

"The smart grid is not really a complex thing from an idea perspective," says Mr. Heimann. "It becomes complex through the incredible amount of pieces that need to work together. It's just a question of time."

## Growth Forecast

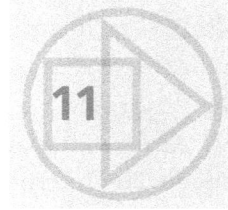

**11**

Without question, the Internet of Things is upon us. The question is not whether it will grow—undoubtedly it will—but where, how, and how fast. As Mary Murphy-Hoye puts it bluntly, "The Internet of Things is going to happen to us, either badly or well."

Enterprises can begin to prepare by thinking through the questions raised in this book. The questions and answers will vary by industry.

In this section, we peer into the murky haze of the future and examine how the IOT may grow in the years ahead.

### Return to the Forum

Let us return briefly to the IRF 2009. At the end of the forum, participants were invited to prognosticate as part of the perennial Crystal Ball Session. Moderator Dan Woods stirred up dissent when he suggested

that the Internet of Things would remain mired in intranets for 10 years and that a true IOT might never be attained. This was a question meant to provoke discussion—and provoke it did. Many rose to the defense of the IOT and insisted that it was closer than skeptics acknowledge.

"The Internet of Things is here," insisted Prof. Khosla of Carnegie Mellon. "I don't know what the argument is."

Ms. Murphy-Hoye of Intel acknowledged that the IOT still lacks some important tools. But she insisted that would not stop it from growing.

"There are things that we're still missing, but it doesn't mean that we can't go quite a distance without having all those things in place," she said. "It doesn't mean we have to wait."

Prof. Günther of Humboldt University echoed the earlier observation that the IOT is going through the typical trough of disillusionment in the hype cycle. He has little doubt that the IOT will come to fruition.

"I think in 20 years we will see that all the major supply chains will be enabled by Internet of Things kinds of technologies," he said. "We'll see that a lot of households, a lot of consumers are using this."

The question is: who will ride this wave and who will be crushed by it?

"A lot of companies will make a lot of money and a lot of companies will lose a lot of money," says Prof. Günther. "The art will be to be on the right side of this."

## Making the Case

One key challenge is to come up with better tools and expertise to identify areas where the potential ROI seems favorable. As Prof. Günther noted, it's hard to pinpoint situations, verticals, industries, or applications where IOT technologies might be profitable.

"Very often managers err on the cautious side," he added. "Rather than spending a lot of money and then finding out after two years that they lost a lot of money and being fired, they prefer to do nothing. If you do nothing and most other people did nothing you're not going to get fired."

Dr. Kaiserswerth of IBM said the business case needs to be made at the top of enterprises. "The sale needs to be made to CEOs," he said. "It's not made to the people further down in the organization, because it's not in their P&L, and so why should they do this? Out of the goodness of their heart? No."

Once again, the participants lamented the lack of business cases. But many argued that the absence of business cases was not as dire as the naysayers had claimed. Indeed, they saw promising examples popping up all over—and more every day. The subsequent research conducted for this book, and the use cases presented in Chapter 10, proved that they were correct.

Many saw clear progress. Prof. Heuser of SAP Research cited one project in which SAP helped design an RFID-based system to shut fire doors at Frankfurt airport—an early deployment which remains in operation today. ""Don't tell me that there are no business cases out there," he said. "It's a matter of being creative and having the right need."

Ms. Alsdorf agreed. She faulted the IT industry for being excessively pessimistic. "The IT industry is making a very, very big mistake at the moment, because they are not really listening to the customers," she said. "At the end of the day the customers will tell us what they really want…We can discuss over and over again and tell each other that there is no clear business case. There are very clear business cases already in the market."

## Driving Forward

What are going to be the major drivers of innovation in the IOT? Cost reduction? Quality? Consumers? Enterprises? According to Prof. Mühlhäuser of Technische Universität Darmstadt, the value drivers are likely to vary by industry.

Prof. Mühlhäuser returns to the auto industry as an example. He sees several potential drivers in the auto sector: cost reduction, enhanced functionality, energy efficiency, or environmental friendliness. Among European automakers there is a strong emphasis on rich features, such as headlights that follow the road around curves. The IOT is an obvious next step toward functionality like in-car measurement systems, computer assistance, or car-to-car infrastructure.

"I see IOT around the corner because we are pretty much at the extreme tension of the screw with many other technology solutions," said Prof. Mühlhäuser.

"As we all know, most of the innovation is happening in the software and computer supported space," he adds. "So it seems to me to be very,

very natural that the next technology step will involve the Internet of Things very heavily."

Prof. Mühlhäuser believes the first waves of IOT technologies will occur in high value niches. Inevitably, these niche applications will confront problems like standardization and scalability. Over time, however, these niche efforts will overcome these barriers, grow, and lay the foundation for more widespread adoption.

He compares the current stage of intranets to the fiefdoms of the Dark Ages. Over time, these closed systems will merge into larger kingdoms and finally one big interconnected mesh.

"Intranets are medieval," he said. "They will crumble like cookies."

## Value Drivers: What Will Push the IOT?

What are other value drivers? Once again, Prof. Fleisch provides trenchant insight about the future of the IOT. He predicts that these technologies will become more widespread as they prove their value. Prof. Fleisch has identified seven main value drivers for IOT technologies:

- **Simplified manual proximity trigger:** Smart things can communicate their "name" (unique identification number) in a fast and convenient way. Examples include self-checkout of library books, access control in buildings payment procedures, or pet tagging. These enable customer self-service of costly tasks and reduce labor costs. Essentially, they allow outsourcing to the customer

- **Automatic proximity trigger:** Some systems may automatically trigger a transaction when objects cross a certain threshold—for example, when a consumer steps out of a store with merchandise that has not been purchased. Another example is a BMW that automatically opens its door when the driver carrying its key approaches. This value driver brings speed, accuracy, and convenience and reductions in labor, failure, and fraud. It also can provide data that can be used to improve these same processes

- **Automatic sensor triggering:** Data collected from sensors can expand upon ID sensing. For example, an olive tree farm might deploy sensors that monitor temperature, moisture, and sunshine and adjust irrigation accordingly

- **Automatic product security:** An object can be secured with a miniature computer that is equipped with some security technology such as cryptography. Its validity can be checked with methods such as challenge-response operations. These techniques are already well-established in ATM cards or car keys. Other applications include proof-of-origin, anti-counterfeiting, product pedigree, and access control. Yet these kinds of applications have relatively expensive requirements for computing resources, power, and digital keys. As a result, they tend to be limited to applications where high values and risks are at stake

- **Simple and direct user feedback:** Smart things can feature simple and small feedback mechanisms. This could be a signal that confirms a process worked, such as the beep of an item scanned at the checkout counter. Or a car key could sense the location of the car and give directions to help the owner find it in a crowded parking lot

- **Extensive user feedback:** Devices may open the door to rich services. A user-friendly computer such as a mobile phone serves as a gateway that links the tagged item with its homepage or any other resources on the Internet. This value driver could have numerous applications: a service might allow a person to scan a bottle of wine and pull up reviews from consumer forums or *The Wine Spectator*. In fact, it could be linked to an endless trove of information—price comparisons, information about the labor standards and political conditions in the country where it was produced, health and allergy warnings, wine reviews from friends in the users social network, and so on

- **Mind-changing feedback:** Sometimes information can change behavior in desirable ways. Imagine a toothbrush that interacts with a comic book character in the bathroom mirror—parents would not have to work so hard to motivate kids to clean their teeth. Similarly, imagine smart meter-based applications that show us our consumption of power or water, compare it to our peers and suggest ways of conservation. Companies could use this information to design new products to appeal to consumer desires. An insurance company might give discounts to customers who agree to put a data recorder in their car. This benefits two sides: the consumer pays less and the company can reconstruct accidents, see driving habits, and attract cautious drivers

## Consumers as Drivers?

What kind of a role will consumers play in the IOT? After all, the consumer space proved to be extremely fertile in the technological sector. Many business technologies like mobile, Web 2.0, and online collaborations and social networking filtered into the enterprise only after they had grown wildly popular in the consumer realm. Will a similar dynamic play out with IOT?

Opinions differ. Dr. Nochta of SAP Research doubts the consumer space will be a major driver of the IOT in the near future. The upfront costs are too steep and only big players like large corporations will be in a position to provide the capital.

"Everybody's thinking today about the smart meters and other home automation technologies and smart devices, but only a few people want to buy a smart meter for their households," Dr. Nochta says. "Everybody expects the utility companies to install those devices. Major development to really show the direction of the entire thing will come out of the enterprise."

But at the IRF 2009, many participants took a different view. In the Crystal Ball discussion, several participants saw the consumer field as an opportunity for explosive growth.

Dr. Skellern of NICTA saw such promise. He cited the example of Google Latitude, which can plot the location of your smart phone to a Google map. It seems clear that Google is going to leverage this service into an opportunity to sell localized advertising. Such capabilities open the door for other services: checking the whereabouts of your kids, seeing if your friends are nearby, and so on. Google has spawned a huge array of small services. Dr. Skellern predicted an explosion of the IOT business cases—with consumers leading the way.

"They're going to spread virally," he said. "The Internet of Things is going to start happening, within the next 12 months. In a big way."

The consumer space will put pressure on the business space. According to Dr. Skellern, consumers will bring these applications into the enterprise through the back door. In a sense, this will be a reprise of the manner that employees brought consumer applications and technologies into the workplace and gave birth to the phenomenon of "shadow IT."

"Pressure is going to lead to companies having to deliver that," he predicted. "It's going to move into the workplace. What you're going to have to deliver will be forced on you by your workers."

Similarly, Prof. Kofman of Telecom ParisTech University predicted that consumer applications will turn out to be a major driver of the IOT. He noted that consumer space drove innovation in Internet and mobile. Who would have guessed that SMS would explode as it did? Now some companies even sell RFID tags that can be personalized with whatever digital information the user desires. "I have the feeling that the innovation will come, as it happened with the Internet, from the end users, with simple applications," he said.

Prof. Kofman predicted that smart devices will seep into all aspects of our lives. He returned to the oft-discussed example of home health care. Perhaps someday homes will be equipped with cameras that detect falls or injuries. He speculated that these systems will have the ability to understand if the person is ill or injured and take appropriate action, such as giving instructions or alerting the hospital.

"The important thing here is that the object is becoming part of our environment," he said. "We are sharing the same space. The rules of living will change somehow, and we will have to start thinking about governance of these things. Maybe these objects will have rights one day.

"The Internet of Things is not about things being connected to the Internet," Prof. Kofman said. "It's about things becoming the Internet."

## Timing?

How quickly will the IOT become a common fixture of our lives? Some talk in time frames of 5 years, others 10, and others 20. Again, the answer likely will vary between areas.

Dr. Ranier Zimmermann expects that it will take 10 years before we see mass adoption of the IOT. As we saw with the Internet, business was slower to catch on than consumers. "The business world is more conservative," Dr. Zimmermann said. "Return on investment plays an enormous role and everybody will be cautious. But as it is convincing, people will do it."

Dr. Zimmerman says issues of infrastructure investment will become much easier to solve when the benefits are no longer so abstract. If Wal-Mart's suppliers had seen clearer benefits, the story would have taken a different turn. "The moment their suppliers see huge economies in using the IOT for their own purposes, they will be much more motivated."

He believes the likely path will be in industries with high value and high risk of something going wrong through spoilage or contamination, like meat production, pharmaceuticals, or fruits and vegetables—all sectors involving "anything that easily goes bust."

Business cases obviously will play an important role in the speed of adoption. As Dr. Zimmermann notes, government mandates also may act as drivers. These need not be requirements to deploy certain technologies *per se*; rather, as governments demand more traceability for products like pharmaceuticals or food, enterprises may see technologies like digital product memory as the best route to compliance. In the course of meeting these requirements, enterprises may realize that they can expand upon these technologies for business reasons. A business case then may become more apparent.

## Services and Things

At the IRF 2009, Prof. Heuser reminded the group that the IOT and Internet of Services are intimately intertwined. "In the end, the Internet of Services and the Internet of Things are really flip sides of the same coin."

Others echoed this theme. This subject is a central part of the research of Dr. Charles Petrie, a senior research scientist at Stanford University and member of SAP's Platform Thought Leadership Council.

Dr. Petrie and his colleague Dr. Christoph Bussler of Merced Systems explored this topic in the essay "The Myth of Open Web Services: The

Rise of the Service Parks."[7] Academics have long dreamed of a free and open Internet where people enjoyed infinite choice and could freely combine services. This vision remains theoretically possible, yet in reality it remains a distant possibility. Business people want trusted services and recognizable brands. Dr. Petrie illustrates the forces at work with an analogy of the cathedral and the bazaar.

In the cathedral, a small elite decides what they think we need and builds a large "closed garden." The provider completely controls what goes into the application. Microsoft is an obvious example of this monolithic cathedral-style approach.

In the bazaar model, many providers offer an array of options on the open Internet and customers can choose what they want and put them all together. Once prognosticators thought the Internet would spawn this sort of open marketplace of competing products and services. But this hasn't turned out as expected. As Dr. Petrie notes, this storyline might apply to software, but not products or web services.

The cathedral model may be losing its hold, but it is not being replaced by the bazaar. Instead, it is being replaced by the branded community. Take the example of books. There simply isn't a thriving bazaar of booksellers who compete on the basis of price. Instead, Amazon is the 900-pound gorilla that dominates the market. Why? People don't want to shop around among unknown providers—just as they don't want to shop in a street market where they don't know which merchant is trustworthy and which is a crook. People prefer to simplify their choices and rely on trusted brands. As Dr. Petrie and Dr. Bussler write, "people want simplicity and quality rather than choices."

According to Dr. Petrie, this shift toward branded communities has major implications for emerging web services technologies, software and services. A few companies are establishing Web service communities. Dr. Petrie and Dr. Bussler compare these communities to industrial service parks.

These new virtual service parks will offer packages of web services with their own sets of rules. Each platform will have sets of business

---

[7] Petrie, Charles and Bussler, Christoph, ""The Myth of Open Web Services: The Rise of the Service Parks," *IEEE Internet Computing*, May/June 2008, *http://www-cdr.stanford.edu/~petrie/online/peer2peer/serviceparks.pdf*

objects and perhaps some semantics (not as sophisticated as semantics developed by academics). They will build unique technologies and protocols for combining services and allow some degree of customization. They will allow external service providers to play in the park if they follow these rules. These parks will offer service-level guarantees and common runtime systems that greatly facilitate implementation of the services. These new models are the next step beyond software as a service: platform as a service.

These parks will offer new capabilities. Customers will have more choice than they do today. New services and processes can be resold within the platform and customers will be able to pick and choose. They will be able to buy packages and customize them to some degree. This model is similar to the vision of the open bazaar, except within a closed environment. There may be some interoperability with parks (semantics will facilitate this quest).

According to Dr. Petrie and Dr. Bussler, these parks will be neither cathedral nor bazaar. Rather, they will be branded communities. They predict a stampede toward these new models and the emergence of a few established service parks operated by popular and trusted brands. They identify several obvious front runners: SAP, Oracle, IBM, Microsoft, Salesforce.com, and Google.

"Service parks signal the death, or at least suspended animation, of the dream of freely combining heterogeneous web services from various providers," they write. "But service parks are also revolutionary themselves."

Sooner or later, these isolated service parks and communities will have to interoperate. Dr. Petrie predicted that semantics would link now-isolated communities.

He remained optimistic, but he cautioned that these developments will take time.

"You just have to put these things in perspective," he said at the IRF Crystal Ball Session. "The technology is here and it's happening. It might take 10 or 20 years."

Dr. Petrie predicted that someday we will have a more advanced version of software installation wizards. These super apps will help with anything—arranging schedules for a touring band, business challenges, or personal projects of private people. At some point, these services will be so customizable, fine grained, and interoperable that individuals will be able to function like OEMS.

"At some point car manufacturers like Daimler will go away," he said. "Daimler might be the best person in the world for you to get your ceramic pistons from, but there'll be somebody else who's really good at processes. All this stuff will be commoditized, and consumers will be able to work with them to order exactly the car they want, and the car will be a lifelong service. This is all part of the long tail. People are going to be empowered to program the world."

But there will be tradeoffs, added Dr. Petrie. This level of interconnections means that we will always have people watching what we are doing. "We won't ever be alone," he says. "We'll be connected to lots of other people, doing lots of things with other people all the time, increasingly. Privacy will just become a cult concern."

## Embedded Systems + Network = IOT

Prof. Wahlster believes the IOT is actually close at hand. He notes that 90% of processors now go into embedded, invisible systems. The question is how to connect all these embedded devices.

"If we network these embedded systems, then we end up with the Internet of Things," says Prof. Wahlster. "I'm optimistic."

Prof. Wahlster was the final expert interviewed for this book and his insights helped put the discussion in perspective, just as his wise observations helped set the stage months before at the conference. He points to two promising developments: the all-IP factory and all-IP car.

In Germany, there has been a strong movement toward the all-IP-based factory. These are based on wifi and Internet technologies. This ushers in a new era of "plug and produce"—the manufacturing equivalent of "plug and play"—in which factory components will be easily interchangeable. "You can just plug it in and by using the IP-based networking services, this new machine is registered in the overall context and you're ready to control it from your manufacturing execution system," he says.

The factory becomes a fully instrumented environment. All the machines can talk to each other. Managers can easily track aberrations on the shop floor and quickly pinpoint problems. According to Prof. Wahlster, semiconductor plants have already pioneered these kinds of advanced instrumented systems over the last few years and now these technologies are spreading into other industries.

"You can much more easily connect the factory to your standard business operations, which are already on the Web," he says. "The SAP stuff, the manufacturing execution system, and the ERP system come much closer together. We even talk about SOA architecture for the factory shop floor, and this is only possible if you move completely to IP-based protocols."

Some naysayers predicted that the all-IP-based factory would never work because the factory still needed a "red button" emergency shutdown. But the IP-based factory has even overcome this barrier and been certified by German authorities. "There are two companies already selling this," he says. "Even the red button in the factory where you can stop all processes is now allowed on an IP-basis," says Prof. Wahlster. "So all these people who criticized the standard and said 'IP is not really secure and safe enough' now have become very quiet."

The all-IP-based factory is an important development because it shows progress toward common standards. Similarly, there is a consortium moving toward the all-IP-based car. According to Prof. Wahlster, BMW has produced a prototype IP car that, instead of having the normal 70 embedded control units, has only four bigger processors.

"I think IP going into the factory and the car is really a success story, and I think other areas will follow soon," he says. "Once we have IP as a basis, then we can step to semantic product memory."

"On the low level, we can have this interconnection between embedded systems," he continues. "But on the higher level, the Internet of Things means that the things understand each other. This can only be done in more narrow domains, like cars talking to each other. It's a long way until the car talks to the pizza delivery."

But progress on semantics, standards, and networking suggest we are moving in the right direction, adds Prof. Wahlster. He expects that it will take another 5 years for a complete breakthrough.

But he remains confident: the IOT is coming. According to Prof. Wahlster, it boils down to a simple equation: embedded systems plus network equals IOT.

"We are already living in a world, especially in the developed countries, of embedded systems everywhere," he says. "The next thing is to network them, and then we're already at the Internet of Things."

## Looking Ahead: The Future Internet

Prof. Wahlster's enthusiasm underscores an important message: we live in exciting times.

We will continue to explore these frontiers in business and technology at the SAP Research International Research Forum. The IRF 2010 will focus on the Future Internet. This idea encompasses many topics familiar to IRF participants, including the Internet of Things, the Internet of Services, network convergence, cloud computing, and Web 2.0. All of these topics have been explored in previous IRFs.

The Future Internet is upon us, but we still struggle to fully grasp it. People view it from different perspectives. Some technologists discuss it in terms of future protocols. Some software people see it in terms of business services like cloud computing or software as a service. Others envision embedded systems or smart grids. In reality, all of these are important pieces.

In the IRF 2010, we will investigate how these elements fit together and how the Future Internet will evolve. Once again, leading minds from business and academia will pool their collective wisdom as they dissect the Future Internet and related issues like business model innovation in the digital economy, technological infrastructure, and the challenges of moving toward this vision.

We look forward to making the 2010 conference, the fifth International Research Forum, the best one yet.

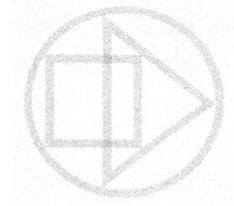

## Authors

**Prof. Dr. Lutz Heuser,** Executive Vice President and Head of SAP Research, is responsible for SAP's overall research portfolio. Prof. Heuser serves on the advisory boards of FhG-Fokus, Berlin and FhG-IPSI, Darmstadt, and is a member of acatech, the German Academy of Science and Engineering. In 2008 and 2009, he served as Chairman for ISTAG, the European Commission's Information Society Technologies Advisory Group. Prof. Heuser is an Adjunct Professor of the Queensland University of Technology in Brisbane, a Professor at the Technical University of Darmstadt, and holds an Honorary Ph.D. from the Technical University of Dresden.

**Henrike Paetz** is Director for Research Dissemination at SAP and has 10 years' experience in communication and change management. Her responsibilities include numerous highly visible events, platforms, and publications presenting technology and innovation topics for various audiences—the IRF being one of them. Together with her team, she supports dissemination of partner communication for about 50 collaborative research projects across all channels and media and oversees the strategic integration of research findings into SAP technology communication. One recent project involved creating a dedicated program for gender research for the European Center for Women and Technology (ECWT).

**Dan Woods,** CTO and Editor of CITO Research, has a background in technology and journalism. As a CTO, Dan built technology for companies ranging from Time Inc. New Media to TheStreet.com. Dan holds an M.S. from Columbia's Graduate School of Journalism and a B.A. in Computer Science from the University of Michigan. Dan has written 20 books about technology, including *Mashup Corporations, Wikis for Dummies, Driven to Perform*, and *In Pursuit of the Perfect Plant* and writes a weekly column for Forbes.com. In 2002, Dan founded Evolved Media, a technology communications and analysis firm serving companies from startups to the Fortune 100. CITO Research, which Dan founded in 2009, is a community-based research organization serving the needs of CIOs, CTOs, and other IT professionals.

## Participants

**Prof. Dr. Karl Aberer** has been a professor of distributed information systems at the Ecole Polytechnique Fédérale de Lausanne in Switzerland since 2000. His research interests are decentralization and self-organization in information systems with applications in peer-to-peer search, Semantic Web, trust management, and mobile and sensor networks. Prof. Aberer received his Ph.D. in mathematics in 1991 from the Swiss Federal Institute of Technology, Zurich. From 1991 to 1992, he was a postdoctoral fellow at the University of California, Berkeley. In 1992, he joined the Integrated Publication and Information Systems Institute of GMD in Germany, where he led the Open Adaptive Information Management Systems research division. Since 2005, he has been director of the Swiss National Research Center for Mobile Information and Communication Systems. He is on the editorial boards of the *VLDB Journal*, the *ACM Transaction on Autonomous* *and Adaptive Systems*, and the *World Wide Web Journal*. He provides consulting on research and science policy to the Swiss government.

**Claudia Alsdorf** has 10 years' experience in executive management, development, licensing and commercialization of new consumer electronic products and services across the domains of the Internet, online commerce exchanges, virtual reality, and wireless. Ms. Alsdorf is CEO and founder of Echtzeit AG, a global provider of 3D and online exchange products and services. In 2002 she joined SAP as Vice President of Communications Development in Global Communications. In 2004 Ms. Alsdorf became Head of SAP Inspire. From 2006 to 2009, she was also responsible for SAP Research Communications. In 2010 she became CEO of Original1, a joint venture between SAP, Nokia, and Giesecke & Devrient that delivers solutions that fight counterfeiting and piracy.

**Rick Bullotta** is the co-founder and Chief Strategy Officer of Burning Sky Software, a pioneer in collaborative, real-world-aware applications. Mr. Bullotta was previously CTO at Invensys Wonderware, a leading provider of manufacturing operations software solutions, and a vice president with SAP Research. Mr. Bullotta was co-founder and CTO of Lighthammer Software Development. At Lighthammer, Bullotta identified and created a new market segment for "manufacturing intelligence and integration" software. Lighthammer was later acquired by SAP. He has also contributed to a number of industry standards organizations including ISA-95/IEC 62264 and the OPC foundation.

**Prof. Dr. Jan Eloff** is Research Director of SAP Meraka UTD/SAP Research CEC Pretoria and a professor in the Computer Science Department at the University of Pretoria. From 2005 to 2008, he chaired the university's School of Information Technology. Prof. Eloff served as president of the South African Institute of Computer Scientists and Information Technologists from 2004 to 2006. In 2006, he received the Outstanding Academic Achiever award and in 2008, he received the Leading Minds in Research award. He is associate editor of *Computers & Security*. At the University of Pretoria, he helped found the Information and Computer Security Architectures research laboratory. He is recognized for his research in computer security by the National Research Foundation (NRF) in South Africa.

**Claudia Funke** is the Director of the Munich office of McKinsey & Company. She leads the company's German High Tech Sector area and is a member of McKinsey's global High Tech Sector leadership group. Ms. Funke works primarily for industry leaders in software, IT-services, and telecommunication, as well as industrial high tech in the enterprise customer segment. Ms. Funke joined McKinsey as a fellow in May 1994, was elected principal in May 2000, and became director in June 2006. Her main areas of industry expertise include Enterprise ICT Services, Clean Tech., and Med Tech.

**Prof. Dr. Hans Gellersen** is a professor of interactive systems in the computing department at Lancaster University. Previously, he was affiliated with the University of Karlsruhe as director of the Telecooperation OfficeComputing Department from 1996 to 2000. Prof. Gellersen was at Karlsruhe's Telematics Institute from 1993 to 1996 and at the University of Kaiserslautern in 1992 as a faculty research assistant. Prof. Gellersen holds both an M.Sc. and a Ph.D. degree from the University of Karlsruhe. His research interest is in ubiquitous computing and human-computer systems. His specific interests are the integration of sensors and perception in interactive systems, interaction with large numbers of networked artifacts, new interaction techniques, and mobile/wearable collaborative applications.

**Michel Guillemet,** a member of the Bull Executive Committee since 2002, is in charge of Research and Development. His work at Bull started in the field of system and processor architecture for mainframes. After successfully managing the development of the first custom CMOS mainframe processor, he is presently responsible for enterprise server development, HPC system development, and Bull's complete R&D activities.

**Prof. Dr. Oliver Günther** is Dean of the School of Business and Economics at Humboldt University in Berlin, Germany. He also directs Humboldt's Institute of Information Systems and its Interdisciplinary Center on Ubiquitous Information. Prof. Günther has held visiting appointments at the European School of Management and Technology, Tsinghua University in Beijing, ENST Paris, UC-Berkeley, the University of Cape Town and, most recently, at SAP Labs in Palo Alto. He is on the Advisory Board of SAP Research and served as an IT strategy consultant and board member to numerous government agencies and high-tech companies. His current research interests include business applications of social networks, RFID architectures, IT productivity, and security and privacy in ubiquitous computing.

**Prof. Dr. Otthein Herzog** has been a chaired professor for artificial intelligence at the University of Bremen, Germany. He directs the Center for Computing and Information Technologies and the Mobile Research Center. He is also CEO of mobile solutions group GmbH. Prof. Herzog's current research interests include wearable and mobile computing, multi-agent systems in logistics settings, semantic concepts for time and space, knowledge discovery, content-based analysis and retrieval methods for images and videos, and digital image and video processing. Before Prof. Herzog joined the University of Bremen, he worked for 16 years at IBM Germany. Herzog has co-authored and co-edited 24 books and more than 160 scientific papers. He is a fellow of the Gesellschaft für Informatik and an ACM member.

**Prof. Dr. Ryo Imura** is a founder, President, and CEO of Mu-Solutions company in Hitachi Ltd., which was established in 2001 to promote the world smallest RFID "Mu-chip." In 2004 he was an Executive Managing Director of Tracing & Tracking RFID Systems Division. Currently, Prof. Imura is Global Business Director in the Information & Telecommunication Systems Group and also Corporate Officer for Hitachi, Ltd. In addition, he is a professor at the University of Tokyo and a guest lecturer for MBA students at the Haas School of Business at the University of California at Berkeley.

**Prof. Dr. Stefan Jähnichen** is head of the Software Engineering Research Group in the Department of Electrical Engineering and Computer Science at the Technical University of Berlin. Since 1998, Prof. Jähnichen has been the managing and scientific director of the Fraunhofer Institute for Computer Architecture and Software Technology. In 2008, he became President of the Society for Information Technology e.V. Prof. Jähnichen received his Ph.D. (Dr.-Ing.) in electrical engineering from the Technical University of Berlin in 1979. Prof. Jähnichen is chief editor of the German research paper *Informatik: Forschung und Entwicklung* and has authored 10 books and approximately 100 journal articles.

**Dr. Matthias Kaiserswerth** has led the IBM research strategy in systems management and compliance since January 2006. In June 2006, he was reappointed Director of the IBM Zurich Research Laboratory. From 2002 until the end of 2005, Dr. Kaiserwerth was the managing director of a large international power and automation company headquartered in Switzerland. In 2000, Dr. Kaiserswerth became director of IBM's Zurich Research Laboratory. Most recently, he worked on smart cards and Java security, which led to the OpenCard industry standard for using smart cards in a Java environment and Visa's Java Card™ Price Breakthrough program based on the IBM Zurich Research JCOP platform.

**Prof. Dr. Pradeep Khosla** received a B. Tech. (Hons.) from the IIT in Kharagpur, India in 1980 and both an M.S. and a doctoral degree from Carnegie Mellon University. At Carnegie Mellon, Prof. Khosla is currently Dean of the College of Engineering. His previous positions include founding director of the Institute for Complex Engineered Systems and program manager at DARPA. Prof. Khosla is a recipient of several awards including the ASEE 1999 George Westinghouse Award for Education, the Siliconindia Leadership Award for Excellence in Academics Technology in 2000, the W. Wallace McDowell Award from IEEE Computer Society in 2001, and the Cyber Education Award from the Business Software Alliance in 2007. He is a fellow of the IEEE, the American Association of Artificial Intelligence, and the American Association for Advancement of Science, as well as a member of the National Academy of Engineering. Prof. Khosla's research has resulted in three books and more than 350 articles.

**Dr. Mathias Kirchmer** is executive partner for Process Excellence at Accenture. He leads the development and delivery of business process lifecycle management offerings, as well as the governance for the management of Accenture's business process reference models and related process assets. Before joining Accenture, Dr. Kirchmer was with IDS Scheer for almost 18 years. Dr. Kirchmer is an affiliated faculty member of the Program for Organizational Dynamics of the University of Pennsylvania as well as a faculty member of the Business School of Widener University, Philadelphia. In 2004, he won a research fellowship from the Japan Society for the Promotion of Science. He is the author of numerous publications, including *High Performance through Process Excellence.*

**Prof. Dr. Daniel Kofman** is the corporate Chief Technology Officer of RAD Data Communications and a professor at the ENST (Telecom ParisTech University) where he heads the Networks, Mobility and Security Lab. For the last 17 years, Prof. Kofman has provided expertise to several companies and public organizations like the European Commission. He is Chairman of the Steering Board of the European Commission-granted Network of Excellence Euro-NF, Anticipating the Network of the Future. Prof. Kofman is a member of the scientific committee of the French parliament. He founded two telecommunication startups and has authored a book on high-speed networking as well as more than 60 research papers.

**Dr. Timo Kosch** is currently a team manager for BMW Group Research and Technology. He is responsible for projects on distributed systems, including cooperative systems for traffic efficiency and active safety, automotive security, and RFID-based vehicular applications, and serves as a coordinator for the European COMeSafety project. He is also currently the head of system development for the German national Car2X field trial. Dr. Kosch studied computer science and economics at Darmstadt University of Technology and at the University of British Columbia in Vancouver. He received his Ph.D. in computer science from the Munich University of Technology.

**Dr. Uwe Kubach** is the Director of the SAP Research Center Dresden. Among other activities the Center is driving SAP's research in the fields of future manufacturing, smart items, and data management and analytics. Dr. Kubach initiated a number of research projects in these domains and regularly acts as an industrial consultant to organizations such as the European Commission. Dr. Kubach has a background in compiler engineering and distributed systems and earned a Ph.D. from the University of Stuttgart and an Executive MBA from the University of Mannheim.

**Prof. Dr. Martti Mäntylä** is research director of the Helsinki Institute for Information Technology. Prof. Mäntylä's present research interests cover a range of topics relevant to future generation communication and computing. His main focus is on user-centered methods for new digital service design but also includes various aspects of mobile digital economy. He is a member of the Finnish Academy of Technology, ACM, IEE, and the Eurographics Association, as well as an editorial board member of the

*Journal of Computing and Information Science in Engineering.* He holds an engineering degree in computer science and operations analysis and a doctoral degree in computer science and applied mathematics from the Helsinki University of Technology.

**Dr. Tiziana Margaria** is chair of service and software engineering at the University of Potsdam, Germany. She received a post-secondary degree in electrical engineering and a Ph.D. degree in computer and systems engineering from the Politecnico di Torino, Italy. Dr. Margaria has broad experience in formal methods for high assurance systems. Her current focus is on formal methods that support reliability and compliance through model-driven service-oriented development. Dr. Margaria is coauthor of more than 100 refereed papers. She is a member of the ACM, IEEE, GI, FME, EAPLS, and EASST. Dr. Margaria is also the ideator and general chair of ISoLA and co-founder of the *International Journal on Software Tools for Technology Transfer,* of the NASA journal *Innovations in Systems and Software Engineering,* and of the *International Journal of Critical Computer-Based Systems.*

**Prof. Dr. Friedemann Mattern** has been a professor of computer science at the Swiss Federal Institute of Technology (ETH) Zurich since 1999. He heads the distributed systems research group and is founding director of the Institute for Pervasive Computing. Prof. Mattern is a member of the editorial board of several scientific journals and book series and has initiated and chaired a number of international conferences, published more than 150 research articles, and edited several books. He is a member of the German Academy of Sciences "Leopoldina," the Heidelberg Academy of Sciences, and of acatech, the German Academy of Technological Sciences. His expertise includes distributed systems and ubiquitous computing, and he is particularly interested in the Internet of Things.

**Dr. Nelson Mattos** is Google's Vice President of Engineering for EMEA. He is responsible for all engineering and product development activities in Google's 12 engineering offices in the region, driving key products and features in areas such as search, mobile, geo, or applications. Prior to joining Google, Dr. Nelson was an IBM distinguished engineer and Vice President of Information and User Technologies. Dr. Nelson was an associate professor at the University of Kaiserslautern, Germany until

1991. He received his Ph.D. in computer science from the University of Kaiserslautern and his bachelor's and master's degrees in computer science from the Federal University of Rio Grande do Sul, Brazil.

**Prof. Dr. Hao Min** is Research Director of Auto-ID Labs and a professor of ASIC & Systems State Key Laboratory at Fudan University. He is also the Chairman and co-founder of Shanghai Quanray Electronics, which he started in 2006. Prof. Min got his Ph.D. from Fudan University in 1991 and then worked in the ASIC & Systems State Key Laboratory. Prof. Min's research areas include VLSI architecture, RF and mixed signal IC design, digital signal processing, and image processing. He has published more than 50 papers in journals and conferences and is the inventor of more than 10 patents (pending).

**Prof. Dr. Max Mühlhäuser** is head of the Telecooperation Lab at Technische Universität Darmstadt, Informatics Dept. The Lab works on smart ubiquitous computing environments for the pervasive Internet. He also heads the RBG division for e-Learning and computing services and is Directorate member of CASED, a center for advanced security research. He is the founder and speaker of a center of research excellence on e-Learning and of a corresponding graduate school funded by the National Funding Agency DFG. Prof. Mühlhäuser has approximately 25 years of experience in areas related to Ubiquitous Computing, Networks and Distributed Systems, and e-Learning. He regularly publishes in journals such as *Pervasive Computing, ACM Multimedia, Pervasive and Mobile Computing, Web Engineering,* and *Distance Learning Technology.*

**Prof. Dr. Wolfram Münch** studied physics, astronomy, and mathematics in Heidelberg, Germany and Cambridge, England. His postgraduate studies focused on turbulence research, and he received his Ph.D. from Cambridge in 1990. From 1990 to 1998, Prof. Münch worked as a research assistant at the Daimler-Chrysler Research Center in Ulm in the area of materials research, focusing on materials for energy management. From 1998 to 2001, he led the Daimler-Chrysler Exchange Group at the DaimlerChrysler headquarters in Stuttgart-Möhringen. In 1991, he held a part-time research position at the Max-Planckinstitute for Solid State Research in Stuttgart, where he worked on ion conduction mechanisms in solid-state bodies. In 2000, he was promoted to professor and became an associate professor at the University of Ulm. Since 2001, he has headed the Research and Innovation division of EnBW, Energie Baden-Württemberg AG.

**Mary Murphy-Hoye,** a Senior Principal Engineer in Intel's Embedded & Communications Group, is a supply chain and information technology solution development expert as well as a trusted advisor across numerous industries—end-to-end retail, industrial automation, high tech, oil & gas, chemical, aerospace and automotive manufacturing, and transportation. Ms. Murphy-Hoye's recent focus has been the creation of Intel's RFID/Wireless Sensor Networks Lab for industry-scale proactive computing experimentation across businesses. She is currently developing multiple embedded Atom research pilots, working on instrumenting the US rail industry, and creating composable collaborative workspaces with Steelcase, a global office furniture leader.

**Dr.-Ing. Bertram Nickolay** studied communications engineering at the University of Applied Sciences of Saarland and continued his studies at the Technical University (TU) Berlin, where he graduated in electrical engineering with a focus on control engineering. In 1990, he completed his Ph.D. at TU Berlin. Dr. Nickolay is the publisher and author of scientific textbooks, as well as publications for image processing and pattern recognition. From time to time, he assumed the position of editor for scientific magazines in the area of sensor technology and pattern recognition. In 1992, he was given the Joseph-von-Fraunhofer Award for his work on learning image analysis systems. He received worldwide popularity through his work on automated virtual reconstruction of torn documents. Due to his initiative, the project for the computer supported reconstruction of torn 'Stasi' (secret service of the former GDR) documents began, which has been regarded as the "largest puzzle of the world."

**Prof. Dr. Victor Nikitin** is Vice Rector and Dean of the College of Business Informatics at the State University – Higher School of Economics, Moscow. His interests focus on federal and corporate information systems, state higher-education governance, innovation in higher education, and IT higher-education standards. From 1994 through 2002, he worked as an enterprise partner unit manager, enterprise unit manager, and account manager with the Moscow office of Microsoft. In 1993, Prof. Nikitin became general manager of the Innovative Center of Russia's Ministry on Architecture and Construction in Moscow. Prof. Nikitin as an engineer with the Moscow State Institute of Electronics and Mathematics Technical University – MIEM. He holds a Ph.D. in cybernetics from the Technical University – MIEM and has attended an MBA course at the European Institute of International Management in Geneva.

**Prof. Dr. Hubert Österle** is Professor of Business Engineering and Director of the Institute of Information Management of the University of St. Gallen, Switzerland, since 1980. He was founder and executive board member of the Information Management Group (IMG AG) from 1989 until 2007. Currently the Editor-in-Chief of *Electronic Markets – The International Journal on Networked Business*, Prof. Österle is also the author of numerous books and other scientific publications and a member of several scientific and industry boards. Prof. Österle's research is focused on corporate data quality, independent living, and sourcing in the financial industry.

**Dr. Charles Petrie** is a senior research scientist working at the Stanford CS Logic Group. His research topics are concurrent engineering, enterprise management, and collective work. Petrie was a founding member of the Technical Staff of the MCC AI Lab, founding editor-in-chief of *IEEE Internet Computing*, and founding executive director of the Stanford Networking Research Center. He is the founding chair of the Semantic Web Services Challenge. He received his Ph.D. in computer science from the University of Texas at Austin.

**Justin Rattner** is Vice President and CTO of Intel. He is also an Intel senior fellow and head of the Corporate Technology Group. In 1989, Mr. Rattner was named Scientist of the Year by *R&D Magazine*. In December 1996, he was featured as Person of the Week by ABC World News for his visionary work on the Department of Energy's ASCI Red System. In 1997, Mr. Rattner was honored as one of the 200 individuals having the greatest impact on the US computer industry. He has received two Intel Achievement Awards for his work in high-performance computing and advanced cluster communication architecture. He is a member of the executive committee of Intel's Research Council and serves as the Intel executive sponsor for Cornell University. Mr. Rattner is also a trustee of the Anita Borg Institute for Women and Technology. Prior to joining Intel, he held positions with Hewlett-Packard and Xerox.

**Runa Sarkar** is an assistant professor in the Industrial and Management Engineering Department of the IIT Kanpur. Ms. Sarkar graduated as a chemical engineer, after which she pursued her master's in environmental engineering at the University of North Carolina at Chapel Hill. A fellow of IIM Calcutta, she specializes in managerial economics and corporate environmental behavior. Ms. Sarkar is an economist interested in socializing the use of ICT to build social capital in rural areas using the power of knowledge networks. She is currently involved in applying social informatics to agriculture. It is her firm belief that ICT deployment using a digital ecosystem approach that protects and nourishes existing social capital would lead to a virtuous cycle of rural development with the lowest possible resource intensity. Her interests lie in sustainable development where business interests are protected in consonance with the furthering of environmental and social objectives. Convinced that there is no free lunch, given the time horizons of business, her efforts in the development sector are aimed at devising mechanisms to make the lunch as cost effective as possible.

**Dr. Joachim Schaper** is SAP Research Vice President of SAP Research EMEA. He oversees SAP's research sites throughout EMEA and is in charge of 6 of SAP Research's programs. Dr. Schaper is also closely involved in the definition of an industrial initiative (European Technology Platform-NESSI) with several strategic research partners, including Atos, BT, Nokia, IBM, Siemens, Thales, Telecom Italia, and Telefonica. He started his career as a system developer with Daimler Benz AG in 1988. He later worked at the European Research Center of Digital Equipment GmbH in Karlsruhe. Dr. Schaper is a member of the MERAKA UTD advisory board and of SFB 627: Nexus. He is also a member of the advisory group for IST (ISTAG), CIP, and various other businesses.

**Prof. Dr. Alexander Schill** is a professor of computer networks at the Dresden University of Technology. His major research interests include distributed systems and middleware, high-performance communication and multimedia, and advanced teleservices, such as teleteaching and teleworking. He holds an M.S. and a Ph.D. in computer science from the University of Karlsruhe, Germany, and received a Dr.h.c. from the Universidad Nacional de Asunción, Paraguay. Prof. Schill also worked with the IBM Thomas J. Watson Research Center in Yorktown Heights, New York and has been involved in various industry collaborations. He is the author of a large number of publications on computer networking, including several books.

**Dr. David Skellern** is an ICT entrepreneur with a background in scientific research, industrial R&D, and engineering education. He has been CEO of the National ICT Australia (NICTA) since mid-2005. Dr. Skellern spent 10 years, designing, building, and commissioning instrumentation and extensions for the Fleurs Synthesis Radiotelescope before joining the academic staff at Sydney University. In 1997, Dr. Skellern co-founded Radiata Inc. to build wireless LAN chips based on research he led at Macquarie University, in collaboration with CSIRO. Radiata demonstrated the world's first IEEE 802.11a chip set in September 2000, and was acquired by Cisco Systems. He was awarded the 2007 CSIRO Tony Benson Award for Individual Achievement in ICT and is an IEEE fellow, an honorary fellow of Engineers Australia, and a fellow of the Australian Academy of Technological Sciences and Engineering.

**Dr. Constant Smits** is Senior Vice President and Chief Software Technology Officer of Philips Healthcare. In this role, Dr. Smits is functionally responsible for software research and development. This responsibility includes overseeing the architecture, technology choices, platforms, methodologies, and tools of Philips Healthcare software activities. Prior to his current position, Dr. Smits was responsible for establishing Philips Healthcare's Informatics. In this role, he was responsible for the Stentor PACS acquisition in July 2005 and the Epic Enterprise IT alliance in November 2003.

**Dr. Guilherme Luís Roehe Vaccaro** graduated from the Universidade Federal do Ro Grande do Sul (UNISINOS) in 1993 with a degree in applied and computer mathematics. He continued his studies at UNISINOS, earning a master's degree in production engineering in 1997 and a doctorate in computer science in 2001. Dr. Vaccaro presently heads the master's program in production engineering and systems at UNISINOS. He has a great deal of experience in the area of production engineering, particularly in operational research and quality engineering. He is focused on research and applied projects on simulation, optimization, knowledge management, and statistical methods.

**Uli van der Meer,** Vice President and General Manager of HP's global Manufacturing and Distribution Industries (MDI) business, has been a member of HP's management team for more than 20 years. He leads the HP MDI Global Industry Vertical, which includes the automotive/aerospace, electronics/semiconductor, oil and gas/utilities, and consumer goods/retail industry segments. Prior to his current role, Mr. van der Meer held various positions in corporate account management, industry sales and marketing, alliance management, and consulting. He earned a master's degree in computer science and economics from the Technical University RWTH Aachen in Germany. He currently is based at the HP Germany headquarters in Boeblingen.

**Sal Visca,** Chief Technology Officer of SAP's Technology Development group, manages the Central Architecture and Applied Innovation CTO Group for NetWeaver and the Business Objects product and technology portfolio. Mr. Visca is responsible for defining architecture guidelines and governance models, technology strategy, and roadmaps, as well as driving a process to bring new technologies and innovations to the product lines. Prior to the acquisition by SAP, Mr. Visca was CTO of Business Objects beginning in 2006. His professional roles included CTO and EVP Engineering of Marqui, a SaaS provider of Content and Communications Management solutions and Co-President and CTO of Infowave Software, a leading provider of wireless computing technology. He also spent more than 12 years with IBM in various senior management and lead architecture positions. Mr. Visca holds an honors B.Sc. degree from the University of Western Ontario in computer science.

**Prof. Dr. Wolfgang Wahlster** is the Director and CEO of DFKI, the German Research Center for Artificial Intelligence, and a professor of computer science at Saarland University. He has published more than 170 technical papers and 8 books on language technology and intelligent user interfaces. Prof. Wahlster is a fellow at Association for the Advancement of Artificial Intelligence (AAAI) and the European Coordinating Committee for Artificial Intelligence (ECCAI). In 2001 he was presented the Future Prize—Germany's highest scientific award presented by the President of Germany. In 2002 Prof. Wahlster was elected full member of the German Academy of Sciences and Literature, Mainz, and was the first German computer scientist elected foreign member of the Royal Swedish Nobel

Prize Academy of Sciences. In 2004 he was elected full member of the German Academy of Natural Scientists Leopoldina, and of acatech, the Council for Engineering Sciences at the Union of the German Academies of Science and Humanities. He serves on the Executive Board of the International Computer Science Institute (ICSI) at Berkeley and the EIT ICT Labs of the European Institute of Innovation and Technology.

**Prof. Dr. Janet Wesson** is Head of the Department of Computer Science and Information Systems at the Nelson Mandela Metropolitan University (NMMU) in Port Elizabeth, South Africa. She is widely published in local and international conferences and journals on user-centered design, user interface design, and usability evaluation. Her current research areas include information visualization, intelligent and adaptive interfaces, and mobile computing. Prof. Wesson is South Africa's national representative on IFIP TC.13 Technical Committee on Human-Computer Interaction (HCI), vice chair of IFIP TC.13, and secretary of WG13.2 (User-Centered Design Methodologies). Prof. Wesson is a South African National Research Foundation-rated researcher and is acknowledged internationally as an expert in HCI education and research.

**Dr. Rainer Zimmermann** graduated in engineering from the Technical University (TU) of Berlin in 1977 and obtained his Dr-Ing in 1985 from the same university. He worked from 1977 to 1985 as a researcher for the TU Berlin and the Fraunhofer Society. He then joined the Commission as a project officer for projects in the fields of production engineering, high-performance computing (1990) and software (1992). In 1995 he became Head of the Unit for Telematics. After heading the Software and Systems Unit (1999) and the Unit for Nanoelectronics and Photonics (2005), he currently heads the Unit for Future Networks of the Directorate General for Information Society and Media at the European Commission.

**Dr. Martina Zitterbart** is a professor in computer science at the University of Karlsruhe (TH). From 1987 to 1995, she was a research assistant in Karlsruhe, receiving her doctoral degree in 1990. From 1991 to 1992, Dr. Zitterbart served as a visiting scientist at the IBM Thomas J. Watson Research Center in Yorktown Heights, NY. She was a visiting professor at the University of Magdeburg and the University of Mannheim and a professor at the Technical University of Braunschweig from 1995 to 2001. Her primary research interests are in the areas of protocols and architectures of communication systems, including the Internet and mobile and sensor networks, as well as networked embedded devices. In addition, Dr. Zitterbart is a member of the IEEE, ACM, and the German Gesellschaft fur Informatik. In 2002, she received the Alcatel SEL research award on technical communication. In 2003, she was general co-chair of the ACM SIGCOMM conference held in Karlsruhe.

## Virtual Participants

**Dr. Ralf Ackermann** holds a Ph.D. in computer science from the TU Darmstadt where he worked in the field of IP telephony services in heterogeneous environments at the Multimedia Communications Lab (KOM). After working as team leader for the ubiquitous computing research group at KOM, and founding a startup company, he joined SAP Research in July 2006. There he worked as a project leader for various research initiatives in the fields of RFID and semantic technologies. In addition, Dr. Ackermann has a strong interest in systems engineering and heterogeneous systems, ranging from microcontrollers and sensor nodes over OS kernel and device driver programming to building large industry grade (communication/real-time) systems. He has a number of referenced publications and works as reviewer for the EU.

**Michael Bechauf** is Vice President of Industry Standards and the SAP Developer Network at SAP AG. He is responsible for managing SAP's participation in industry standards through a company-wide governance process that balances customer benefits with intellectual property risks. Mr. Bechauf serves in several industry activities, such as the Eclipse Foundation Board of Directors, as well as the Java Community Process Executive Committee. He is also managing the SAP Developer Network (SDN), an online collaboration platform for developers in the SAP eco-system with more than 1.9 million members, as well as the SAP Mentor Initiative, a high-impact influencer group consisting of SDN developers, technologists, and bloggers. He has been with SAP since 1997 and lives in the San Francisco Bay Area.

**Prof. Dr. Elgar Fleisch** is Professor of Information and Technology Management, ETH Zurich and University of St. Gallen (HSG). His research focuses on the economic impacts and infrastructures of ubiquitous computing. In the Auto-ID Lab he and his team have developed, in concert with a global network of universities, an infrastructure for the Internet of Things. In the Bits-to-Energy Lab, which he co-chairs with Prof. Mattern of ETH Zurich, he investigates and designs technologies and applications to save electrical power and water. In the Insurance-Lab, Prof. Fleisch, together with Dr. Ackermann of HSG, drives technology-based innovation in the insurance industry. All research projects are joint efforts of industry and academia and Prof. Fleisch's findings have been published in more than 200 scientific journals and books.

Prof. Fleisch is a co-founder of several university spin-offs (for example, Intellion, Synesix, Coguan, Dacuda, and Amphiro), and serves as a member of multiple management boards, and academic steering committees.

**Dr. Christian Floerkemeier** received his Bachelor and Master of Engineering degrees in electrical and information science from Cambridge University in 1999 and his Ph.D. from ETH Zurich, Switzerland, in computer science in 2006. He is currently Associate Director of the Auto-ID Lab at the Massachusetts Institute of Technology. Before joining the Auto-ID Lab at MIT, Dr. Floerkemeier was Associate Director of the Swiss Auto-ID Lab at ETH Zurich. From 1999 until 2001, he worked as head of software development for Ubiworks, an Amsterdam-based software company. Dr. Floerkemeier is the co-founder of the leading open source RFID project Fosstrak and is a member of the EPCglobal Architecture Review Committee. He was the Technical Program Chair of the first international Internet of Things Conference and IEEE RFID 2009, and has published numerous papers on radio frequency identification and pervasive computing.

**Dr. Gregor Hackenbroich** is a research program manager in data management and analytics at SAP Research. His main interests include the management of structured and unstructured data, data integration, and business intelligence. Dr. Hackenbroich and his team developed the Auto-Mapping Core that is used in SAP Business Process Management. He has been responsible for the acquisition and management of several large research projects, including the lighthouse project THESEUS. He received his habilitation in theoretical physics from Essen University, and his doctoral degree in physics from the University of Munich. He has been a Heisenberg Fellow and has won research fellowships from the Alexander-von-Humboldt Foundation and the Studienstiftung des Deutschen Volkes. Dr. Hackenbroich has published more than 50 research papers on computer science and theorectial physics.

**Klaus Heimann,** Senior Vice President within SAP AG's Business Unit Global Industry Solutions, is responsible for SAP's software solutions in Service Industries (media, telco, utilities and waste and recycling). From January 2002 through April 2007, Mr. Heimann headed up the Industry Business Unit (IBU) Utilities at the headquarters of SAP AG in Walldorf, and was responsible for SAP for Utilities, SAP's market-leading business platform for the global utilities industry. Over his 32 years of experience in information technology, Mr. Heimann concentrated on the development and implementation of standard software solutions for various service industries, particularly in the utilities industry. Before joining SAP, he worked for 16 years at a German software company, where, as General Manager and a co-owner for 11 years, he focused on building standard software solutions in customer relations and billing for the European utilities industry. Mr. Heimann studied information technology at the University of Karlsruhe, Germany, graduating with a Diplom-Informatiker.

**Prof. Dr. Marc Langheinrich** is an assistant professor of informatics at the Università della Svizzera italiana (USI) in Lugano, Switzerland. He received a master's degree in computer science from the University of Bielefeld, Germany, in 1997, and earned his Ph.D. from the ETH Zurich, Switzerland, in 2005. Before joining USI in September 2008, he worked as a researcher at the ETH Zurich, at NEC Research in Tokyo, Japan, and at the University of Washington in Seattle. Prof. Langheinrich has published extensively on security and privacy issues in ubiquitous computing, RFID, and the Internet of Things. He has been the program co-chair of both the 5th Intl. Conference on Pervasive Computing (Pervasive 2007) and the First Intl. Conference on the Internet of Things (IOT2008). Prof. Langheinrich is also the general co-chair of the 11th Intl. Conference on Ubiquitous Computing (Ubicomp 2010).

**Dr. Volkmar Lotz** is the Research Program Manager for Security and Trust at SAP Research. His responsibilities include definition and implementation of SAP's security research agenda and the maintenance of a global research partner network. His group's research topics include service security, business process security, secure software engineering, trusted collaboration, compliance enforcement, and adaptive security. Before joining SAP, he headed the Formal Methods in Security Analysis group at Siemens, focusing on security requirements engineering, evaluation and certification, cryptographic protocol verification, and mobile code security. Dr. Lotz is the main contributor to the LKW model, a security model for smartcard processors. His experience includes context-aware mobile systems, legally binding agent transactions, and authorization and delegation in mobile code systems.

**Burkhard Neidecker-Lutz** is the CTO of SAP Research, SAP AG's worldwide research organization since 2003. He joined SAP in 1999. From 1988 to 1998 he worked at the European Applied Research Center of Digital Equipment Corporation in Karlsruhe, Germany. Mr. Neidecker-Lutz is a professional member of the ACM and holds a masters degree in computer science from the University of Karlsruhe.

**Dr. Zoltán Nochta** is Deputy Director of the SAP Research Campus-based Engineering Center (CEC) located in Karlsruhe, Germany. At CEC, he manages teams that work on multiple research projects including smart items, IT security, and privacy protection, as well as smart grids and future energy markets. Dr. Nochta is also the SAP Research Intrapreneur for the Internet of Things and drives an initiative to help position the company as a service provider in this emerging business. He has was also responsible for preparing the content and discussion topics for the IRF 2009 in Dresden. Dr. Nochta received his Ph.D. at the University of Karlsruhe where he conducted research in the area of secure access control and cryptography in distributed systems.

**Dr. Jochen Rode** has a German diploma in business and information technology, and earned a master's and Ph.D. in computer science. His background ranges from networking and software engineering to web application development, human computer interaction, and manufacturing. Dr. Rode joined SAP in 2005 as a senior researcher. In his current position as Development Architect and Smart Items Research Program Manager, his primary interest is applying smart items technologies (such as RFID, sensors, and embedded systems) for SAP's customers. Dr. Rode also leads SAP Research's R&D for the manufacturing domain and manages the

SAP Research Future Factory Living Lab *(www.sap.com/futurefactory)* at Dresden. In this context he leads a number of projects related to device integration, machine data acquisition, and control.

**Dr. Harald Vogt** has been a researcher at SAP Research since 2006. His main research interests are distributed systems, pervasive computing, and software engineering. Currently, he is active in developing systems for the operation of smart electricity grids and electric vehicles. His results will help to prepare SAP's industry solutions to meet future requirements in the utility sector. Before joining SAP and pursuing his Ph.D. in secure communications for wireless sensor networks, he spent a few years at the Darmstadt research center of Deutsche Telekom AG, where he investigated the security of Java smartcards. Dr. Vogt holds a master's degree in computer science from the University of Ulm, Germany, and a Ph.D. in computer science from ETH Zürich.

www.ingramcontent.com/pod-product-compliance
Lightning Source LLC
Chambersburg PA
CBHW021555210326
41599CB00010B/451